岩土与地下工程专业英语

秦哲　王旌　张骞　编

English for Geotechnical and Underground Engineering

清华大学出版社
北　京

内容简介

本书针对岩土与地下工程的专业特点，在完成基础英语学习和部分专业基础课的基础上，着重讲授专业词汇、术语及相关文章的翻译和阅读。内容主要包括专业英语基础、专业英语的文献检索、专业英语的翻译，以及基坑工程、隧道工程、井巷工程、边坡工程、地铁工程、防灾与减灾等相关专业内容。

本书可供土木工程、岩土工程、城市地下空间工程等专业的本科生学习专业英语使用，也可供相关专业的研究人员参考阅读。

图书在版编目(CIP)数据

岩土与地下工程专业英语/秦哲，王旌，张骞编. —北京：清华大学出版社，2022.6
ISBN 978-7-302-59525-0

Ⅰ.①岩… Ⅱ.①秦… ②王… ③张… Ⅲ.①土木工程－英语－高等学校－教材 ②地下工程－英语－高等学校－教材 Ⅳ.①TU4 ②TU94

中国版本图书馆 CIP 数据核字(2021)第 230605 号

责任编辑：袁 琦
封面设计：何凤霞
责任校对：赵丽敏
责任印制：丛怀宇

出版发行：清华大学出版社
网 址：http://www.tup.com.cn，http://www.wqbook.com
地 址：北京清华大学学研大厦 A 座　　邮 编：100084
社 总 机：010-83470000　　邮 购：010-62786544
投稿与读者服务：010-62776969，c-service@tup.tsinghua.edu.cn
质量反馈：010-62772015，zhiliang@tup.tsinghua.edu.cn
印 装 者：三河市科茂嘉荣印务有限公司
经 销：全国新华书店
开 本：185mm×260mm　　印 张：9.25　　字 数：223 千字
版 次：2022 年 6 月第 1 版　　印 次：2022 年 6 月第1 次印刷
定 价：38.00 元

产品编号：090342-01

编 委 会

前　言

近年来，随着“一带一路”倡议的实施，大量基础设施工程兴建，岩土与地下工程迎来了蓬勃发展期。岩土与地下工程专业英语是读者获取专业知识和掌握学科国内外发展动态的一个窗口，也是提升专业英语写作和表达能力的一个工具。希望能为我国岩土与地下工程的安全建设助力。

《岩土与地下工程专业英语》作为高等院校专业英语的系列教材之一，采用中英文对照形式，重点讲述不同类型岩土与地下工程的基本介绍、基础知识、前沿知识和专业词汇。

本书是通过结合编委会的专业领域知识与多年来专业英语课程的教学经验，查阅大量文献形成的一本教材，专业性较强。希望能帮助读者更好地了解本专业的国内外前沿信息，对专业学习起到一定的帮助。

本书语言撰写规范，涉及岩土与地下工程的重要内容。全书共分为8章：第1章介绍岩土与地下工程专业英语的基础，第2章介绍了基坑工程，第3章介绍了地基工程，第4章介绍了隧道工程，第5章介绍了凿井与掘巷工程，第6章介绍了边坡工程，第7章介绍了城区地下空间的常见工程，第8章介绍了常见自然灾害的防灾与减灾。

本书得到了“山东科技大学优秀教学团队建设计划”的资助，由山东科技大学、山东大学、石家庄铁道大学联合编写。编写过程中得到了清华大学出版社的关心和支持，向编写本书的同事以及参考文献的作者表示诚挚的谢意。

由于时间匆忙，水平有限，本书存在的问题及不足之处恳请广大读者给予批评、指教。

编　者

2021年12月

目 录

专业英语基础 (Fundamentals of Specialized English)

1.1 专业英语特点(Features of Specialized English)

1.1.1 语言特点(Language features)

(1) 语言简练,表达明确,不重润饰。

Example 1: The yield criterion for a material is a mathematical description of the combinations of stresses which would cause yield of the material. In other words it is a relationship between applied stresses and strength.

材料的屈服准则指可能导致材料屈服的应力组合的数学表达式。换句话说,它表示的是应力与强度之间的关系。

(2) 逻辑严谨,要领明确,关系清晰。

Example 2: The term "resilience" refers to "the ability to prepare for and adapt to changing conditions and withstand and recover rapidly from disruptions", two important aspects are reflected in the concept of engineering resilience.

术语"复原力"是指"做好准备适应不断变化的条件,承受混乱中断并迅速恢复正常的能力",工程复原力的这一概念反映出两个重要方面。

1.1.2 语法特点(Grammatical features)

(1) 非人称的句式和客观的态度,常用 It 结构、名词化结构。

Example 1: It is necessary to choose a suitable excavation method during construction.

Example 2: It is estimated that in China, two-thirds of the lining tunnels with disease have cracks.

(2) 较多采用被动语态。

Example 3: Before any tunnel can be dug, a geological survey must be made.

使用时注意:主要信息置于句首;尽量避免使用第一或第二人称。

(3) 大量使用非限定动词,即不定式、动名词、现在分词和过去分词。

Example 4：The higher the strength of concrete，the more likely it is to build the more stable structure.

Example 5：The requirement for complex conditions，coupled with the limitation of nondestructive testing of structures，led to a large number of monitoring methods unable to solve a corresponding set of special problems.

Example 6：a）Archimedes first discovered the principle that water is displaced by solid bodies.

b）Archimedes first discovered the principle of displacement of water by solid bodies.

解释：名词化结构，一方面简化了同位语从句，另一方面强调 displacement 这一事实。

Example 7：a）TBM construction method is in the hard rock environment which uses full-section tunneling machine to carry out tunnel construction.

b）TBM construction method is in the hard rock environment using full-section tunneling machine to carry out tunnel construction.

解释：①在科技文中，一般将 how、wh-词引导的从句转换成不定式短语，使句子结构简洁紧凑。②在科技英语中，不常使用非限制定语从句，通常采用非限定动词，即非谓语动词。③以分词短语代替定语从句或状语从句；以分词独立结构代替状语从句或并列分句；以不定式短语代替各种从句；以介词＋动名词短语代替定语从句或状语从句。

（4）较多使用祈使语气和公式化表达，常用 assume that、suppose that、let 句式。

Example 8：Assume that A is equal to B.

（5）条件句较多。

Example 9：If tunnel water seepage can not be treated in time，and it will gradually become tunnel water gusher，It is difficult to treat them.

Example 10：If substituting Eq.(5)into Eq. (8)，we obtain $F=xz$.

（6）长句较多，但结构一般较清晰。

Example 11：In fact，according to the characteristics of tunnel operation，diseases have existed in many tunnels before they were put into operation，and the causes are very complex，which brings great difficulties to the disease detection.

（7）省略句较多。

Example 12：If no well-supported，the probability of tunnel accident may be increased.

通常省略状语从句的主语和谓语、定语从句的关联词、从句的助动词等。

1.1.3 词汇特点(Lexical features)

专业英语词汇分为 3 类：①纯专业词汇；②半专业词汇；③非专业词汇。

（1）较多使用词性转换。

例如：emphasis（名词）和 emphasize（动词），lateral（形容词）和 the lateral（名词）；strong（形容词）和 strengthen（动词）。

（2）词缀和词根。

例如：dis-，un-，il-，kilo-，fore-，-ics，-ward。

(3) 缩写、数学符号及其表达式。

Fig.(figure)

Eq.(equation)

m/s(meter/second)

in.(inch)

ISO(International Organization for Standardization)

1.1.4 结构特点(Structural features)

上述语言、语法和词汇特点都属于"语域分析"的内容,形成了专业英语的基础。为了把握文章要点和重点,提高阅读理解能力。还需进一步了解段落及文章层面的结构特点,以及语言运用中隐含的逻辑思维过程,一般来说,一个自然段中,总有一句来概括段落大意,对理解本段意思至关重要,通常位于段首或段末,很少出现在段中。应予足够重视。另外,还应注意各自然段间的逻辑关系,这对理解也有很大帮助。

1.1.5 专业词汇(Professional vocabulary)

civil engineer:土木工程。

tunnel and underground engineering:隧道及地下工程。

geotechnical engineering:岩土工程。

soil mechanics:土力学。

dead load:恒载。

live load:活载。

critical design load:临界设计荷载。

safety factor:安全系数。

fatigue:疲劳。

subsoil:下层土,底土,天然地基。

settlement:沉陷,沉降。

existing structures:既有结构。

concrete cover:混凝土保护层。

crack width:裂纹宽度。

1. 常用数字的英语读法(English reading frequently used numerals)

1/2:one half,a half,one over two。

1/3:a third,one third,one over three。

1/10:a tenth,one tenth。

1/4:a quarter,one quarter,one fourth,one over four。

2/5:two fifths,two over five。

3/4:three quarters,three fourths。

$4\frac{1}{2}$:four and a half。

$5\frac{5}{8}$：five and five-eighths。

$112\frac{3}{5}$：one hundred twelve and three over five。

134/459：one hundred and thirty four over four hundred and fifty nine。

0.6：zero point six，o point six，naught point six，point six。

0.06：zero point zero six，o point o six，naught point naught six，point naught six。

0.56：zero point five six，o point five six，naught point five six，point five six。

$0.\dot{5}$：zero point five recurring，point five recurring。

$2.2\dot{6}\dot{7}$：two point two six seven，six seven recurring。

6 %：six percent。

6 ‰：six per mille。

2/8 %：two eighths percent，two eighths of one percent。

1/3 m：one third of a meter。

3/4 km：three quarters of a kilometer。

6.5 m/s：six point five meters per second。

13℃：thirteen degrees Sentigrade。

23 F：twenty three degrees Fahrenheit。

20kN：twenty kilonewton。

10mm：ten millimeters。

2. 常用数学符号的读法(English reading for frequently-used mathematical symbols)

+：plus。

÷：divided by。

−：minus。

±：plus or minus 。

×：multiplied by，times。

=：is equal to，equals，is。

≡：is identically equal to，identically equals 。

≈：approximately equal to，approximately equals。

$|x|$：the absolute value of x。

$\bar{x}$：x bar，the mean value of x。

b'：b prime。

b''：b double prime，b second prime，b two primes。

b_1：b subscript one，b sub 1。

b_2：b superscript two，b super 2。

$\dot{x}$：x dot。

$\ddot{x}$：x two dots。

∇：del。

∇^n：n th del。

∞：infinity。

$f(x)$：function f of x。

dx：dee x，differential x。

$\frac{d^n y}{dx^n}$：the n th derivative of y with respect to x。

$\int$：integral。

$\int_a^b x$：integral between limits a and b。

$(a+b)$：bracket a plus b bracket closed。

$a:b$：the ratio of a to b。

x^2：x square，x squared，the square of x，the second power of x，x to the second power。

x^3：x cube，x cubed，the cube of x，the third power of x，x to the third power。

$\frac{\partial x}{\partial y}$：the partial derivative of y with respect to x。

$\sqrt{x}$：the square root of x。

$\sqrt[3]{x^2}$：the third root of x square。

$\log_n x$：log x to the base n。

$x!$：factorial x。

1.2 专业英语应用(Application of Specialized English)

1.2.1 专业文献检索(Professional literature retrieval)

若想了解某一主题的大致内容和涉及范围，那就应该到这类信息集中的地方查找。目前有许多站点，例如虚拟图书馆(WWW Virtual Library，图 1.1)、雅虎信息站(Yahoo)等，将其所收集的资料以主题形式整理。以下展示一些虚拟图书馆的使用案例。

如果要查找某一特定信息，就需要利用关键词来搜索。具有按关键词搜索功能的网络应用程序称为搜索引擎(searching engine)。利用该程序的基本步骤：①在浏览器中键入拥有搜索引擎的主页 URL；②在主页的输入栏键入关键词；③提交查询；④分析查询结果(一般为带有超链接的文本信息，例如相关网站的地址和简要说明)。Internet 上提供搜索引擎的网站非常多。本文以 ScienceDirect 为例(图 1.2)。

Internet 上重要的信息就是各种图书馆电子目录(library electronic catalogs)，它也是 Internet 上最早的资源。目前建立了世界上大约几万个图书馆的电子目录，这些目录源涵盖大学、政府部门、社会团体等组织的图书馆。信息类型多种多样，不仅有大量的全文档电子资料，例如书籍、论文、简讯、公告，而且还有多媒体数据。图 1.3 是几所大学图书馆提供

The WWW Virtual Library : en · es · fr · zh

The WWW Virtual Library

Quick search:

Agriculture
Irrigation, Livestock, Poultry Science, ...

The Arts
Art History, Classical Music, Theatre and Drama, ...

Business and Economics
Finance, Marketing, Transportation, ...

Communications and Media
Broadcasters, Publishers, Telecommunications, ...

Computing and Computer Science
Artificial Intelligence, Cryptography, Logic Programming, ...

Education
Primary, Secondary, Tertiary, ...

Engineering
Architecture, Electrical, Mechanical, ...

Information and Libraries
Information Quality, Knowledge Management, Libraries, ...

International Affairs
International Relations and Security, Sustainable Development, ...

Law
Arbitration, Forensic Toxicology, Legal History, ...

Natural Sciences and Mathematics
Biosciences, Earth Science, Medicine and Health, Physics, ...

Recreation
Gardening, Recreation and Games, Sport, ...

Regional Studies
African, Asian, Latin American, European, ...

Social and Behavioural Sciences
Anthropology, Archaeology, Population and Development Studies, ...

Society

The WWW Virtual Library
Engineering

Quick search:

Narrower Category : Architecture

Acoustics, Vibrations and Signal Processing	Acoustics, Vibrations and Signal Processing Resources pertaining to the physics of sound, noise control and soundproofing. **this resource in English is indexed under:** Engineering, Physics, Sound.
Biotechnology	Biotechnology This directory contains over 3000 links to companies, research institutes, universities, so information and other directories specific to biotechnology, pharmaceutical development fields. It places emphasis on product development and the delivery of products and servi **this resource in English is indexed under:** Biosciences, Engineering.
Chemical Engineering	Chemical Engineering This subject catalog lists information resources relevant to Chemical and Process Enginee **this resource in English is indexed under:** Engineering, Natural Sciences and Mathematics.
Computer Aided Design (CAD)	Computer Aided Design (CAD) This section of Virtual Library is devoted to CAD education. This section contains links to CAD resources associated with the CAD industry. **this resource in English is indexed under:** Computing and Computer Science, Engineering.

图 1.1 虚拟图书馆

(Figure 1.1 WWW Virtual Library)

的电子目录和出版物查询检索页面。

1.2.2 专业英语翻译(Specialized English translation)

翻译的定义：一种语言活动，把一种语言文字的意义用另一种语言文字表达出来。

专业英语的翻译：翻译的一种，在词汇、语法、修辞等方面有专业语言文字的特点。要求译者在英语、汉语和专业知识三方面都具有良好的素质和修养。

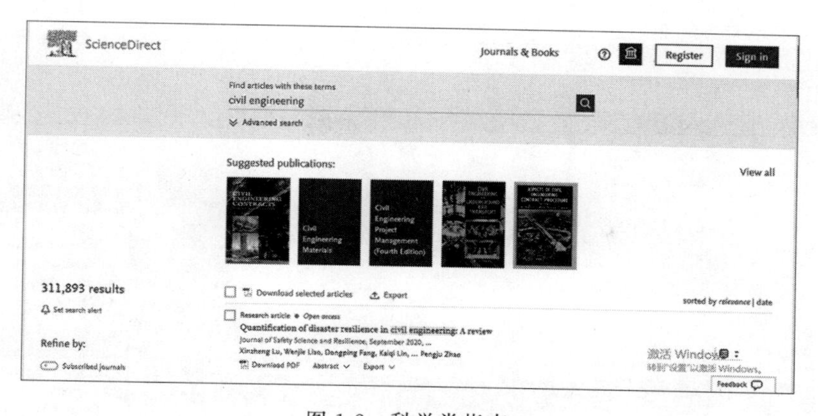

图 1.2 科学类指南

(Figure 1.2 ScienceDirect)

(a)

(b)

图 1.3 图书馆类目录

(Figure 1.3 Libraries Catalog)

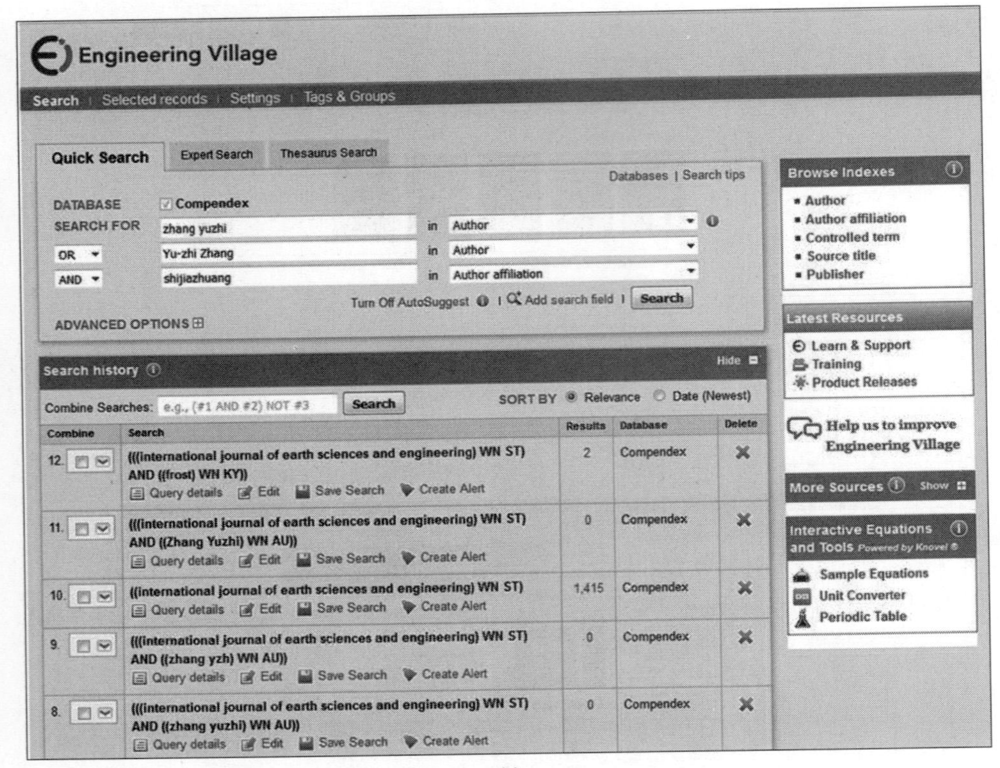

(c)

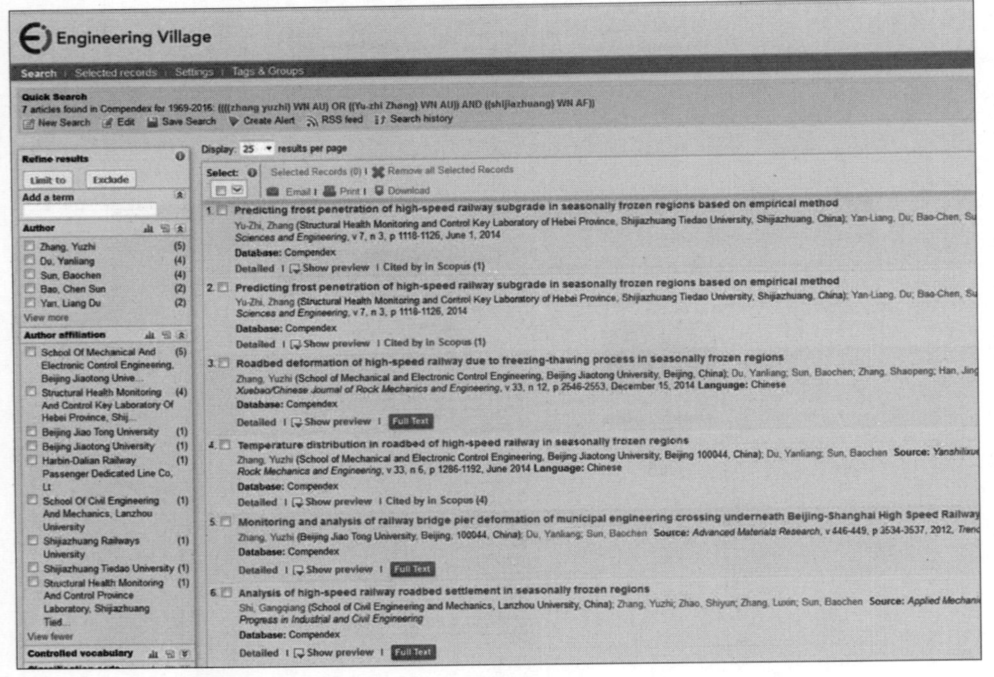

(d)

图 1.3 （续）

（Figure 1.3 （Continue））

1. 概述(summarization)

1) 翻译的标准

Example 1: The importance of foundation stability can not be overestimated in the economic development.

译文 1: *在经济的发展中,地基稳定性的重要性再怎么估计也不过分。*

译文 2: *在经济发展中,地基稳定性的重要性无论怎么强调也不过分。*

Example 2: A novel solution to characteristics of soil under dynamic load has become popular in North America although not yet in Europe.

译文 1: *一种解决动荷载作用下土壤特性的新方法已经在北美流行起来,但在欧洲还没有。*

译文 2: *对于土在动荷载作用下的特性,已经有了一种新的解决办法,这种办法在北美已流行起来,然而在欧洲却未能做到这一点。*

Example 3: The redistribution of base pressure usually follows the failure of tension between the base and the foundation.

译文 1: *基底压力的重新分布通常伴随着基底和地基之间张力的破坏。*

译文 2: *在基底压力重新分布时,基底和地基之间的拉力通常会被破坏。*

翻译的标准是什么? 翻译的标准是信、达、雅。信: 忠实原作,忌无依据揣猜;达: 通顺,流畅;雅: 优雅,得体。

专业英语应体现科技内容的科学性、逻辑性、正确性和严谨性,故专业英语翻译更着重于"信"和"达"。

2) 翻译的过程

(1) 阅读理解: 正确理解原文的词汇含义、句法结构和习惯用法;要准确理解原文的逻辑关系。

(2) 汉语表达: 使用汉语的语言形式恰如其分地表达原作的内容。

(3) 检查校核: 理解和表达不是一次就能圆满完成,要反复琢磨,逐步深入。

Example 4: According to the possible displacement direction of the structure, the earth pressure can be divided into"active earth pressure"or"passive earth pressure".

按结构物可能位移方向的情况,土压力可分为"主动土压力"和"被动土压力"两种。

此处,earth 不能译为地球。必须从上下文联系中去理解词义,从专业内容上去判断词义。

Example 5: Action is equal to reaction, but it acts in a contrary direction.

译文 1: *作用力等于反作用力,但作用方向相反。*

译文 2: *作用力与反作用力大小相等,但作用的方向相反。*

译文 3: *作用力和反作用力大小相等,方向相反。*

Example 6: With the increase of load, the scope of shear failure gradually expands, and finally forms a continuous sliding surface in the soil, and loses stability.

随着荷载的增加,剪切破坏的范围逐渐扩大,最终在土体中形成连续的滑动面,并丧失稳定性。

2. 英汉语言对比(contrast between English and Chinese)

1) 词汇对比

词汇对比主要分为3个方面：①词义；②词的搭配；③词序。

a) 词义

(1) 英汉词汇意义一一对应：civil engineering，flexible pavement，foundation。

(2) 英语词义比汉语广泛：material，machine，reduce。

(3) 汉语词义比英语广泛：soil mass，university，rock soil。

(4) 部分对应，两者的意思有彼此不能覆盖的部分：beat，do，state。

b) 词的搭配

reduce speed	减小速度
reduce to powder	粉碎
reduce the temperature	降低温度
reduce the time	缩短时间
reduce construction expense	削减施工开支
reduce the scale of construction	缩小施工规模
reduce the numbers of engineering accidents	减少工程事故数

Example 7：First，curing measures should be selected which will give long service life and good foundation.

首先，应当选择能延长使用寿命和获得优质地基的养护措施。

Example 8：Two or more sensors can also be operated together to improve performance or system reliability.

也能同时操作两台或更多传感器，以改善性能或提高系统可靠性。

c) 词序

(1) 定语的位置。

Example 9：The effective stress in soil refers to the intergranular stress transmitted by soil particles，which controls the volume and strength of soil.

土壤中有效应力是指土粒所传递的粒间应力，以此控制土壤的体积和强度。

(2) 状语的位置。

Example 10：The forces keeping the basics straight must，by a fundamental law of statics，equal the load tending to fold it up.

根据静力学基本定律，保持各基础呈直线的力必等于易于将其折叠起来的荷载。

Example 11：To the extent possible，the foundation concrete is placed keeping the excavation dry.

尽可能在保持基坑干燥的条件下灌筑基础混凝土。

2) 句法对比

a) 句子结构

(1) 英语简单句转换成汉语复合句结构。

(2) 英语复合句结构转换成汉语简单句。

(3) 英语复合句结构转换为各种汉语复合句结构。

(4) 英语的倒装句转换为汉语正常语序。

(5) 英语被动结构转换为汉语主动结构，反之亦然。

Example 12: Considered from this point of view, the question will be of great importance.

从这一视角考虑，此问题就十分重要。(英语简单句→汉语复合句)

Example 13: In a simple form, because the Swedish slice method neglects the forces on both sides of the soil strip, it can not meet all the equilibrium conditions, so the calculated stability safety factor is low.

简言之，瑞典条分法因忽略了土条两侧各种作用力，不能满足全部平衡条件，故算出的稳定性安全因数偏低。(英语简单句→汉语因果句)

Example 14: Water power stations are always built where there are very great falls.

水力发电站总是建在落差很大的地方。(英语状语从句→汉语简单句)

Example 15: It is essential that civil engineering students have a good knowledge of rock and soil mechanics.

学土木工程的学生必须掌握岩土体力学知识。(英语主语从句→汉语简单句)

Example 16: Cell phones, which have many uses, cannot carry out creative work.

手机虽有很多用处，但不能进行创造性的工作。(英语主从复合句→汉语转折偏正复合句)

Example 17: Then comes the analysis of internal forces.

接下来进行内力分析。(英语倒装句→汉语正常语序句)

Example 18: Soil mechanics and soil stabilization techniques have been used in the construction of foundation stability and foundation pit excavation.

在地基稳定及基坑开挖的施工中，业已采用土力学和土壤稳固技术。

b) 句序

(1) 时间顺序。

英语中表示时间的从句可置于主句之前或之后，顺序灵活；汉语中则先发生的先说，后发生的后说，且时间通常置于句首。

Example 19: One must inevitably touch upon the technical aspects when discussing the stability of soil slope and foundation.

在讨论土坡和地基稳定性时，不可规避地必然涉及其中的技术方面。

(2) 逻辑顺序。

英语复合句表示因果关系或条件与结果关系时，叙述顺序比较灵活，原因或条件从句置于主句前后均可，而汉语中大多是原因或条件在前，结果在后。

Example 20: This time no one was killed or injured in the landslide accident, for great attention was paid to safety.

由于十分重视安全问题，这次滑坡事故中无人伤亡。

3. 翻译的基本方法(basic methods of translation)

主要分为以下4种。

(1) 直译：尽量按原文词语和词性译出。

(2) 转译：①词义转换；②词性转换；③句子成分转换。

(3) 省略：冠词、代词、动词、介词、连词等的省略。

(4) 增补：句法、意义、修辞等需求予以增补。

1) 直译

Example 21：Internet is of great help to our work.

因特网对我们的工作大有帮助。

Example 22：Soil slope refers to soil mass with inclined slope.

土坡是指具有倾斜斜面的土体。

2) 转译

为了使译文合乎汉语语法和修辞习惯，经常需将原文某些词句加以转换，包括对词义、词类、语序、句子成分、句子结构和表达方式等的转换。

a) 词义转换

(1) 引申。

Example 23：Constant—and variable-water-head method are indoor methods to determine the permeability coefficient of soil.

定水头法和变水头法都是确定土壤渗透系数的室内方法。

Example 24：Rock direct shear is 1.1m wide and 1.8m deep.

岩石直剪仪宽1.1m，高1.8m。

(2) 具象化。

Example 25：There are many things that should be considered in the construction of the project.

在项目施工中，有许多事项均宜予以考虑。

(3) 抽象化。

Example 26：We have progressed a long way from the early days of triaxial compression test.

与早期的三轴压缩试验相比，我们已经有了很大的进步。

b) 词性转换

(1) 转换成动词。

Example 27：Rigid foundation is made of brick，block stone，rubble，plain concrete and other materials.

刚性地基是用砖、块石、碎石、素混凝土及其他材料打下的。

Example 28：Hong Kong-Zhuhai-Macao Bridge was built in 2018，it is located in the waters of Lingdingyang in the Pearl River Estuary of Guangdong Province，it is 55 kilometers.

港珠澳大桥建成于2018年，位于广东省珠江口伶仃洋海域内，全长55千米。

Example 29: Although there are many reasons for tunnel leakage, the main reason is geology.

尽管影响隧道发生渗漏水的原因很多,但最主要原因还是地质情况。

Example 30: They are confident that they will be able to build the palaces of Mexico in a short time.

他们坚信短期内就能够建成这里的墨西哥宫殿群。

Example 31: An exhibition of new geotechnical test is on there.

那里在举办新型岩土试验的展览会。

Example 32: The construction of the dam was three months behind.

这座大坝的施工工期延误了3个月。

(2) 转换成名词。

Example 33: The main characteristic of static earth pressure is that the soil behind the wall is in elastic equilibrium.

静态土压力的主要特征是墙后的土体处于弹性平衡状态。

Example 34: These breaking characteristics, however, must be closely watched, for they are constantly being attacked by unfavorable environments.

然而,由于一直受到不利环境因素的侵蚀,所以这些破碎特征必须加以密切观察。

Example 35: Of those stresses the former is gravity stress and the latter is additional stress.

这两种应力中,前者是重力应力,后者是附加应力。

Example 36: Drawings must be dimensionally correct.

制图的尺寸必须正确无误。

Example 37: This page shows the diagram of the self-weight stress of the dam itself.

这一页上有坝身自重应力的图示。

(3) 转换成形容词。

Example 38: There are many ways to apply pressure to structures.

对结构施加压力的方式多种多样。

Example 39: This experiment is an absolute necessity in determining the permeability coefficient.

测定通透系数时,这种实验是绝对必要的。

Example 40: It is a fact that no geotechnical structural materials is perfectly elastic.

事实上,没有哪种岩土结构材料是完美弹性体。

(4) 转换成副词。

Example 41: Engineers have made a careful study of the properties of these new basics.

工程师们已仔细研究了这些新型基础的性质。

Example 42: We find difficulty in simulate the experiment indoors.

我们发觉室内模拟这种试验很难。

(5) 句子成分转换。

Example 43：Attempts were made to find out ways for reducing construction expenses.

曾试图找出削减施工开支的办法。(主语转换成谓语)

Example 44：The test results are in good agreement with whose obtained by theoretical analysis.

这些试验结果与理论分析所得结果十分吻合。(表语转换成谓语)

3) 省略

(1) 省略冠词。

Example 45：The soil mass with a stable safety factor greater than one is stable.

稳定安全因数大于1的土体是稳定的。

Example 46：The column is the pivotal part of a structure.

柱是结构中的中枢部分。

(2) 省略代词。

Example 47：If you know the internal forces，you can determine the length of the component.

如果已知内力，就能确定构件长度。

(3) 省略动词。

Example 48：All kinds of testers perform basically similar functions but appear in a variety of forms.

一切测试仪的功能都基本相同，但形式丰富多彩。

Example 49：The flow chart shown in Fig.1 intended to illustrate the programming process.

图1所示的流程图旨在说明编程过程。

(4) 省略介词。

Example 50：Most material expand on heating and contract on cooling.

大多数材料热胀冷缩。

(5) 省略连词

Example 51：In downstream areas where the river passes through a broad gentle flood plain，geotechnical engineers may be asked to build flood protection works.

在下游地区，河流流经广阔平缓的洪泛平原，会请求岩土工程师们去修建防洪工程。

(6) 省略含义冗余的词。

Example 52：Triaxial compression test can be used to determine the shear strength of silt.

三轴压缩测试能测定粉土的剪切强度。

4) 增补

(1) 根据句法上的需要。

Example 53：Hence the reason why effect of vibration liquefaction of soil is so often viewed with suspicion by the owner.

因此，这正是业主屡屡以怀疑的眼光看待土壤振动液化效果的原因。

(2) 根据意义上的需要。

Example 54: the first production　　*第一批产品*

the experts　　*专家们*

Example 55: Oxidation will make steel structure rusty.

氧化作用会使钢结构生锈。

Example 56: Testing is a tedious process and enough patience is required for its mastery.

进行测试是一个冗长过程,要求有足够耐心方能掌握。

Example 57: The compressibility of soil increases with the increase of volume compression coefficient.

土的压缩性随体积压缩系数的增加而增加。

Example 58: A designer must have a good foundation in rock mechanics and soil mechanics.

设计人员必须在岩石力学、土力学这两个方面有良好的基础。

(3) 根据修辞上的需要。

Example 59: The question is really a material itself rather than a instrument problem.

这一问题确实与材料本身有关,而不是某一仪器问题。

Example 60: The instability of the foundation may be caused by the existence of soft soil layer in the foundation and the inclined rock layer or fracture zone under the soil layer.

地基不稳定可由地基中存在软土层,以及土层下面有倾斜岩层或断裂带两者所导致。

(4) 重复上文出现过的词。

Example 61: Another test method is to apply horizontal shear stress to the specimen. If the corresponding displacement increases continuously, the specimen is considered to have been sheared.

另一种测试法是对试样施加水平剪应力,若相应位移不断增加,则认为该试样已被剪切。

Example 62: A synthetic material equal to that alloy in strength has been produced, which is very useful in geotechnical engineering.

一种在强度上和那种合金相当的合成材料已经生产出来了,这种合成材料在岩土工程中十分有用。

4. 特殊句型的翻译(translation of special sentence patterns)

1) 被动句型

a) 译成汉语主动句

(1) 原文中的主语仍译作主语,将被动语态的谓语译成:加以……,是……的。

Example 63: The slope of foundation pit excavation is determined by the nature of soil and the depth of excavation.

基坑开挖的坡度是由其土壤性质和开挖深度确定的。

(2) 原文中主语在译文中作宾语,译成汉语无主语句。

Example 64: Attempts are also being made to produce subway tracks with more strength and durability, and with a lighter weight.

目前仍在尝试生产强度更高、耐久性更好,而且重量更轻的地下轨道。

(3) 用英语句中的施动者作汉语句中的主语。

Example 65: The size of the rails was made more precise by forging or rolling.

采用锻造或轧制的方式使道轨的尺寸更加精密。

(4) 将英语句中的某一适当成分译成汉语句中的主语。

Example 66: Much progress has been made in geotechnical engineering in less than one century.

不到一个世纪,岩土工程学即取得了重大进展。

b) 译成汉语被动句

Example 67: The engineering proposal is predicted accurately by numerical simulation.

通过数值模拟使工程提案得以准确预测。

c) 译成独立结构

把原句中的被动语态谓语动词分离出来,译成独立结构。

Example 68: It is believed that the construction waste is blamed for such problems as occupying land, affecting city appearance, destroying vegetation, water, air and soil pollution.

有人认为施工垃圾造成一系列问题,例如侵占土地,影响市容,破坏植被,以及污染水体、大气和土壤。

这种译法常用于一些固定句型,类似结构还有:

It is asserted that... 有人主张……

It is suggested that... 有人建议……

It is stressed that... 有人强调说……

It is generally considered that... 普遍认为……

It is told that... 有人曾经说……

It is well known that... 众所周知……

有时,对某些固定句型翻译时不加主语,例如:

It is hoped that... 希望……

It is supposed that... 据推测……

It is said that... 据说……

It must be admitted that... 必须承认……

It must be pointed that... 必须指出……

It will be seen from this that... 由此可见……

2) 否定句型

a) 部分否定

英语中 all, both, every, each 等词与 not 搭配时,表示部分否定。

Example 69: All these basics materials are not good products.

这些基础材料并非都是优质产品。

类似结构有 not...much；not...many；not...often；not...always。

b）意义否定

Example 70：The geological environment is too complex for us to survey accurately.

地质环境过于复杂，难以精确勘察。

常见的词组有 but for，in the dark，free from，safe from，short of，far from，in vain，but that，make light of，fail to 等。

c）双重否定

Example 71：There is no foundation without more or less settlement under the action of load.

在荷载的作用下，没有一种地基不或多或少发生沉降。

not ...until；never ... without；no ... without；not(none) ... the less；not a little。

3）强调句型

（1）It is（was)＋被强调部分＋that（which，who）...。

Example 72：It is these factors which need to be considered and which have led to the complexity of rock mass stability study.

正因为这些因素需要加以考虑，才导致岩体稳定性研究的复杂性。

Example 73：It is this kind of support system that the foundation pit engineering needs most urgently.

基坑工程中最急需的正是这种支护体系。

（2）It is（was）not until＋时间状语＋that...。

Example 74：It was not until 2014 that a great new tunnel was built through the Guanjiao mountain in Qinghai.

直到2014年才在青海建成一条贯穿关角山的大型隧道。

（3）在强调句中，被强调的可以是一个词或词组，也可以是一个状语从句。

Example 75：It is not until the mortar bolt can be drilled and grouted properly to obtain the designed strength in the foundation pit that the soil nailing wall can be fully utilized to resist earth pressure.

只有做到在基坑中对砂浆锚杆正确钻孔注浆并使之达到设计强度时，方能充分利用土钉墙来抵抗土压力。

5. 长句的翻译（translation of long sentences）

1）顺译

（1）主谓之间的逗号切断。

Example 76：The main reason for placing multiple anti-slide piles on the sliding surface is to make the sliding degree of the slope within a reasonable range，but especially to pay attention to the relative displacement from one anti-slide pile to the next is not too large.

在滑移面上布设多根抗滑桩的主要原因，就是让边坡的滑移度处于合理范围之内，但要特别注意紧邻桩之间的相对位移不易过大。

(2) 在并列或转折处切断。

Example 77: Jointly or separately graduate school's faculty and graduate students research the issues and make recommendations to the graduate school in time.

研究生院各位导师与研究生或联合或分别研究此类问题,并及时向研究生院提出建议。

(3) 在从句间加逗号或分号。

Example 78: In the course of designing antislide piles on the slope, you have to take into account what kind of load antislide piles will be subjected to; where the load is going to be; the pile spacing and the cross-sectional shape of the antislide pile and whether the row of antislide piles is single or double.

在设计边坡上抗滑桩进程中必须考虑:抗滑桩将承受何种荷载;荷载作用在何处;桩间距以及抗滑桩的截面形状;抗滑桩选单排还是双排。

2) 倒译

在翻译时,将英语长句全部或部分倒装。

(1) 全部倒装。

Example 79: Master's degree students only account for 0.62% of the national population, although the number of master's degree students is increasing year by year.

虽然硕士研究生人数逐年增多,但仅占全国人口的0.62%。

(2) 部分倒装。

Example 80: It is extremely essential that the patent specifications should describe the technical field which designed by patent, background of the technology, the detailed content of the invention, a description with drawings and the specific implementation method.

对于专利所针对的技术领域、该技术背景、该发明的详细内容、附图说明以及具体实现方法等项,专利说明书中均宜予以描述,对这几点的描述极为重要。

3) 拆译

为汉译行文方便,可将原文某一短语或从句先行单独译出,再利用适当的概括性词语或通过语法手段将其同主语联在一起,重新组成全句。

Example 81: Soil arch effect can be divided into horizontal soil arch, which is the soil arch on the horizontal surface caused by soil deformation between piles, and vertical soil arch, is the soil arch which causes shear damage to the soil above the sliding surface and prevents the soil from sinking.

根据空间位置可将土拱效应分为横向土拱和竖向土拱。水平土拱是由桩间土体变形所引起的水平面上的土拱,竖向土拱是对滑动面以上土体造成剪切损伤并防止此土体下沉的土拱。

Example 82: The overall test program for virus content was developed by hundreds of engineers in the United States is a combination of the various test parts of the virus with the help of production, transportation and testing departments, by conducting a total content test across the entire team.

美国数百个工程师研发的病毒含量总体测试方案，是通过跨整个团队实施的总含量测试，即在生产、运输和检测三部门协助下，把对病毒的各种测试部分组合为一体。

在翻译时应先理清定语从句与先行词的逻辑关系。

(1) 译成前置定语。

Example 83：College English is usually the only subject that college students are never excused，whatever major they choose.

不管是选修何种专业，院校英语通常是学生们唯一不能免修的科目。

(2) 译成并列句。

Example 84：Their main interest is inbedding rock slopes，where they may reduce shearing strength between rock and soil mass of slope.

它们主要关注于顺层岩质边坡，此处可减少斜坡岩石与土体之间的剪应力。

(3) 译成状语从句。

Example 85：This is particularly important in cohesive soil because the permeability of this fine-grained soils is too weak and mechanical properties vary with water content.

在黏合土中这一点尤其重要，因为这种土颗粒细，透水性太弱，力学性质随含水量而变化。

(4) 译成单句中的一部分。

Example 86：These landslide geological parameters simulated by software must be considered when designing a stable anti-slide pile.

通过软件模拟的这些滑坡地质参数，在设计稳定的抗滑桩时必须予以考虑。

Example 87：Tunnel meaning of water gushing is that there is a certain pressure head of water out.

隧道涌水的含义是，有一定压头的水往外冒。

6. 有关数量的翻译(translation about quantity)

1) 成倍增加

表示数量成倍增加的句型较多，常用的有：

A is N times as large (long，heavy ...) as B；

A is N times larger (longer，heavier ...) as B；

A is larger (longer，heavier ...) than B by N times。

Example 88：The speed in China is 3 times faster high-speed rail as compared to ordinary train.

中国高铁的速度是普通列车的 3 倍。

Example 89：If the load length is doubled，keep the uniform load unchanged，the bending moment becomes four times as great.

若保持均布荷载不变而荷载长度加倍，则弯矩为原来的 4 倍。

Example 90：This kind of support can shorten the construction time by more than two times.

这种支护方式能使施工时间缩短一半以上。

2）成倍减少

表示成倍减少含义时，通常包含以下句子成分：

reduce by N times，reduce N times as much (many) as；

reduce N times，reduce by a factor of N；

reduce to N times，reduce N-fold；

N-fold reduction，N times less than。

Example 91：The time cost has reduced two thirds.

时间成本减少了 2/3。

Example 92：The advantage of the present scheme lies in a twofold reduction in cast.

目前这一方案的优点在于削减成本 1/2。

3）不确定数量

表示不确定数量的常用词语有 about，around，some，nearly，roughly，approximately，or so，more or less，in the vicinity of，in the neighborhood of，a matter of，of the order of 等。

A weight around 26 tons	26 吨左右的重量
200km or so	大约 200km
A force of the order of 50kN	50kN 量级的力

表示不确定数量的词组有 teens of，tens of，decades of，dozens of，scores of，hundreds of，thousands of。

Example 93：Soil nailing wall：the composite formed by the reinforced bar in the slope and the soil around it，and the supporting structure similar to the gravity retaining wall composed of the surface layer.

土钉墙：由坡体中的加筋杆件与其周围土体，以及面层所构成的类似重力挡土墙的支护结构所形成的复合体。

Example 94：Anti slide pile：in order to support the sliding force of the sliding mass and stabilize the slope，it is a pile structure through the landslide mass and deep into the sliding bed.

抗滑桩：是一种用以支挡滑体的滑动力并稳定边坡，穿过滑坡体深深插入滑床的桩柱结构。

Example 95：Box foundation：the closed box is composed of reinforced concrete base plate，top plate，side wall and a certain number of internal partition walls.

箱型基础：由钢筋混凝土的底板、顶板、侧墙及一定数目的内隔墙构成的封闭箱体。

Example 96：Prestressed anchor cable：the prestressed method is used to anchor the anchor cable against the anchor head into the rock mass through the hole of the weak structural plane of the rock mass，and connect the sliding body with the stable rock mass，which is used to strengthen the cable support of the slope.

预应力锚索：采取预应力方法把锚索靠锚头通过岩体弱结构面的孔洞锚入岩体，把滑体与稳固岩体层联在一起，用于加固边坡的索状支撑。

1.2.3　撰写科技论文(Scientific & eechnological papers)

1. 论文内容(paper content)

科技论文在内容上主要包括：①科技论文的结构；②论文各基本组成部分的写作；③时态及语态；④词汇及语法。

2. 论文体例(paper format)

国际标准化组织(ISO,我国为成员国)科技论文的组成部分包括：

(1) 标题及副标题(title and subtitle)；

(2) 作者(author(s))及所在单位和地址(institution and address)；

(3) 摘要(abstract)；

(4) 关键词或主题词(keywords or subjects)；

(5) 引言(introduction)；

(6) 材料和方法(materials and methods)；

(7) 结果(results)；

(8) 讨论(discussion)；

(9) 结论(conclusion)；

(10) 致谢(acknowledgement)；

(11) 参考文献(references)。

针对论文各基本组成部分的写作,这里将对几个主要组成部分进行介绍。

1) 题名

标题一般不采用句子,而采用名词、名词词组(短语)的形式,通常省略冠词。

从内容上,在拟定论文标题时应注意以下几点：

(1) 恰如其分而又不过于笼统地体现论文的主题和内涵。

(2) 词语选择要规范,要便于二次编制文献题录、索引、关键词等。

(3) 尽量使用名词性短语,字数控制在两行之内。

Example 1：Viscoelastic Foundation

黏弹性地基

Example 2：Bayesian Technique for Evaluation of Material Strengths in Existing Structures

用于评价现行结构的材料强度的贝叶斯技术

Example 3：Numerical Geotechnics

数值岩土力学

2) 署名

署名一般紧跟在论文标题之后,包括作者单位、通讯地址、职称、学历等信息,也有置于论文第一页页脚的。

BRIDGE RELIABILITY EVALUATION USING LOAD TESTS

By Andrzej S. Nowak[1] and T. Tharmabala[2]

[1] Assoc. Prof. of Civ. Engrg., Univ. of Michigan, Ann Arbor, MI 48109

[2] Res. Ofcr., Ministry of Transp. and Communications, Downsview, Ontario, Canada M3M1J8

3) 摘要

摘要是论文的梗概，是对论文的扼要描述。摘要应提供论文的主要概念和所讨论的主要问题，使读者从中可获得作者的主要研究活动、研究方法和主要结果及结论。

摘要一般分为两类：信息性的和指示性的。当今绝大多数科技期刊和会议论文要求作者提供信息性摘要。

(1) 信息性摘要(informative abstract)。主要报道论文的研究目的、方法、结果及结论。通常应提供尽可能多的定量或定性信息，能定量表达的一定要全部量化，要充分反映论文的创新之处。

(2) 描述性摘要(descriptive abstract)。主要介绍论文的论题，或者概括表述研究的目的。无须介绍方法、结果、结论的具体内容，也无须用数据定量描述。

a) 摘要的要素

研究目的 Objective or Purpose。

研究过程及采用方法 Process and Methods。

主要结果或发现 Results。

主要结论和推论 Conclusions。

b) 摘要的特点

一般来说，较短的论文摘要不超过正文的5%，较长的论文约占3%。另外不同期刊对摘要长度有不同的要求。语言上必须精练简洁，尽量避免列举例证等可在正文中陈述的细节。行文无须口语化，尽量避免非通用缩略语，勿采用主观语气的第一人称。不宜只是对标题的简单扩展。论文每一论点都要凸显出来。摘要是全文的缩影，需简短明了，即便未读全文也能大体知晓论文的主要信息。

c) 摘要中的常用句型

Example 4: This paper describes the objects, contents, significance and impact of Information Superhighway project being constructed.

Example 5: The main purpose of this paper is to contribute to the development of more rational system reliability-based structural design and evaluation specifications。

Example 6: The method is based on a radial-space division technique in conjunction with automatic generation of unit center-line vectors.

Example 7: Results of numerical examples indicate that the proposed method has good accuracy with the Monte-Carlo simulation method.

4) IMRaD 结构

撰写英文科技论文的目的是参与国际交流，第一步得构思结构。最简单有效的是采用IMRaD (Introduction, Materials and Methods, Results, and Discussion)结构，此种论文具有简明清晰、逻辑性强等特点，因而广为采用。

3. 论文结构(paper structure)

1) IMRaD 结构的逻辑性

研究的是什么问题？对应 Introduction 部分。问题是怎么研究的？对应 Materials and Methods 部分。发现了什么？对应 Results 部分。发现意味着什么？对应 Discussion 部分。

2) 引言(Introduction)

引言的作用是提供足够的研究背景和内容，使读者领悟作者研究工作的构想与布局思路。撰写引言时，作者要通过研究背景介绍，指出目前研究中存在的不足，引出值得研究的问题或现象及论文的主题，最后点出研究的意义。

引言的写作原则有以下几个：

(1) 介绍作者的研究领域，叙述研究概况与进展，提供背景资料，指明引用的经典、重要、说服力强的文献。

(2) 介绍评论国内外其他学者对该问题或现象曾经发表的相关研究，即文献回顾。

(3) 指出仍有某个问题或现象值得进一步研究，点明此次研究的意义。

(4) 介绍作者的研究目的或研究活动，突出本研究的重要结果和发现，避免自我评价。

撰写引言时，应注意下列问题：

(1) 避免内容空泛或过于简略，力求如实反映作者的创新；

(2) 不过多强调作者过往的成就；

(3) 引言不宜重述摘要和解释摘要；

(4) 引言无须对实验的理论、方法和结果作详尽叙述；

(5) 引言不提前引用结论和建议；

(6) 在引言介绍本人的工作时，不宜过分评价；

(7) 引言可不编章节号，也可编号“0”。

3) 方法(Methods)

方法部分主要用来描述所用的材料、实验装置、实验方法、理论模型、计算方法。写好这部分的关键在于提供恰到好处的细节，避免过于简单或烦琐(太繁复或不必需的公式、推导可放入附录)。这一部分中不需要涉及结果。

注意事项：

(1) 清楚、翔实地描述实验材料和方法；

(2) 重点描述新的研究或方法；

(3) 按照实验先后顺序来排列所介绍的内容；

(4) 注意语态应用。

4) 结果(Results)

结果是指作者通过实验观测或调查研究所得出的结论以及各种图像和数据。表达时要客观真实，简明扼要，准确无误。在结果部分如实地汇报结果或数据即可，无须加入自己的解释，让结果和数据来体现研究结论。在利用图表表达结果时，图表应该简明、清晰、准确，还应附有扼要说明。

5) 讨论(Discussion)

在讨论部分，要回答引言中所提的问题，评估研究结果所蕴含的意义，用结果去阐释问题的答案。因为其中包含了作者的观点和解释，所以在行文时需注意语气，不可夸张；同时注意避免无关紧要或不相关联的内容。讨论分析是论文的精髓所在，内容可包括：①提炼原理，揭示关联，进行归纳；②提出分析、模型或理论；③解释结果与作者的分析，以及模型或理论之间的联系。

6）结论（Conclusions）

结论是在理论分析和实验的基础上，通过作者的概括、总结、推理而对实验结果进行的一种创造性、指导性和经验性描述。应完整、准确、鲜明地表达作者的观点，是对整篇论文的概括总结。在结论部分，作者应阐明论文的主要结果及其重要性，同时点明局限性或有所保留之处。注意以下几点：

（1）措辞要严谨，不能编造无法从论文里导出的结论；

（2）语气要坚决，不能模棱两可；

（3）尽量做到言简意赅；

（4）不要做自我评价。

7）参考文献（References）

参考文献的主要作用是反映论文的科学依据，表现作者对他人研究成果的尊重，向读者提供文中引用有关资料的出处，或因节约篇幅或叙述方便，在论文中虽涉及而没有展开的有关内容来提供的详尽文本。

参考文献的内容可包括书籍、期刊文章、学位论文、会议资料、电子文献资源等，一般要包含作者姓名、文献名称和出版信息等内容。

4. 时态及语态（tense and voice）

1）时态（tense）

科技论文中常用的时态有一般现在时、一般过去时、一般将来时以及现在完成时。按照惯例，有些时态选择可与作者所要传达的信息内容相对应。

一般现在时：不受时间限制的理论，普遍真理。

一般过去时：涉及科技史，强调某一结论基于在此之前发生的某一事实。

现在完成时：一种是类似于一般过去时的使用；另一种是强调某动作持续一段时间。

一般将来时：研究设想。例如：

Example 8：Perhaps future studies will be able to connect more precisely with the data.

Example 9：A three-dimensional finite-element model will be made in this paper, which will give more detailed analysis of this phenomenon.

2）语态（voice）

对无须或难以说明的施动者的情况，就用被动语态。科技论文主要说明事实，一般用被动语态较多。但也不是绝对如此。譬如："A exceeds B"读起来要好于"B is exceeded by A"。通篇采用被动语态会让读者弄不明白是在引用别人的工作，还是作者自己的工作。因而具体使用哪种语态取决于所要强调的重点，同时应考虑表达的简练和精确。

5. 词汇及语法（vocabulary and grammar）

1）形容词和副词（adjectives and adverbs）

在科技论文中使用形容词和副词要慎重。像 fairly，quite，rather，several，very，

somewhat, much, amazing, mainly, briefly 这样的修饰词在科技论文中最好不用。爱希比教授认为,"this very important point"不及"this important point"来得简洁客观,而"this point"则更佳。另外,论文中能用形容词做定语的,尽量不用名词来做。比如用"experimental results"而不用"experiment results"。

2）冠词(articles)

中文作者对英文冠词掌握得通常不够好,例如经常会忘记定冠词"the",克服的方法之一是检查名词,如果名词前没有加不定冠词"a"或"an",而该名词又非抽象名词或不可数名词,则要考虑是否要加上定冠词"the"。下面总结了一些使用定冠词的规则以供参照。

第二次提及：We proposed a new model. The model is ...。

最高级：The most important parameter。

序数词：The first slide。

特指：The only research in this field。

通用知识/独有事物：The Government/the moon。

带后置定语,例如 of：短语 The behavior of the soil。

3）分词(participles)

中文作者在写英文论文时常常会写出这样的句子：

After closing the incision, the animal was placed in arestraining cage.

Having completed the study, the bacteria were of no further interest.

从语法上分析,这两个句子的隐含主语分别为："the animal"和"the bacteria",但作者省略掉的真正的主语其实是"the experimenter"。所以这两个句子都造成了歧义。这是典型的使用分词从句造成的错误。

如果改用一般从句,这种错误就避免了：

After the incision had been closed, the animal was placed in a restraining cage.

Once the study was completed, the bacteria were of no further interest.

4）报告/评估动词(reporting / evaluative verbs)

典型的报告动词有：Others found, reported, showed, proposed, used, developed, studied, suggested, observed, described, discussed, demonstrated, presented, gave。

主要用来表达作者的论点、看法或主张,比如：The author argues / contends / maintains / believes / claims (by the writer does not totally agree; the writer is not convinced) / implies (suggests sth. without saying it directly)。

使用合适的报告动词可以有效地总结作者的观点：The author suggests / points out / notes / stresses / emphasizes / pays particular attention to / adds / goes on to say / begins by (+-ing verb) / concludes that (+noun phrase)。

以上是撰写英文科技论文的基本常识,在这里有必要指出：在科技论文中避免使用抱歉词句。譬如："Unfortunately, there was insufficient time to complete the last set of tests."只能让读者认为作者计划不好,懒惰,能力不够。写一篇好的英文科技论文需要反复推敲,修改。写完初稿后最好先搁置一边,两三天后再拿出来修改。定稿前要仔细检查格式、标点符号,核对参考文献,等等,哪一样细节都不应忽视。

基坑工程
(Foundation Pit Engineering)

2.1 地质勘查(Site Investigation)

A thorough and comprehensive site investigation is an essential preliminary to the design and construction of a civil engineering project. The size and type of project will influence the scope of an investigation, but not its necessity: even the smallest job warrants some form of site investigation (Figure 2.1).

深入全面的现场勘查是土木工程项目和施工必不可少的预备步骤。项目的规模和类型会影响勘查的范围,但不影响其必要性:即使是最小的作业也授权某种形式的现场调查(图 2.1)。

图 2.1 工程地质勘查场景
(Figure 2.1 Engineering geological survey site picture)

2.1.1 地质勘查阶段(Geological exploration stage)

A site investigation in connection with a moderate to large project will normally consist of six phases:

与中大型工程项目相关的地质勘查工作,通常分为 6 个阶段:

(1) Preliminary investigation: essentially a desk study, including collection of maps, drawings, details of existing or historic development, local authority information; provision of overall site-selection data, first drawings, brief for next phase.

(1) 初步勘查：主要是准备工作,包括：收集地图、图纸、现有或过往开发详细、本地当局信息;以及提供总体选址数据、首批图纸、下一阶段概要。

(2) General site survey and examination: first examination of site by several specialists, e. g. geologist, land surveyor, soils engineer; preparation of overall site layout, topography, basic geology, details of access; investigation of local conditions such as climate, stream flows, groundwater conditions(generally); provision of brief for next phase after consultation within design team.

(2) 一般现场勘查和检查：由若干专家(例如地质、测量、岩土工程师)进行首次现场检查;准备场地总体布局、地形、基性地质、通道详情;调查当地状况,例如气候、水流、地下水(大体)状况;在设计团队内部协商后,提供下一阶段的概要。

(3) Detailed site exploration and sampling (Figure 2.2): investigation of detailed geology and sub-surface soil conditions using surface surveys, trial pits, headings, boreholes, soundings, geophysical methods, as appropriate; survey of groundwater conditions over a significant period of time; examination of existing and adjacent structures for signs of cracking or settlement; location of underground structures or cavities, buried pipes is, services, etc.; provision of samples for further examination and laboratory testing.

(3) 详细的现场勘探和取样(图 2.2)：酌情采用地表勘查、试坑、标高、钻孔、探测、地球物理等方法,查清地质和地下土壤的详细情况;查清某一重要时段的地下水状况;检查现有及邻近构筑物有无开裂或沉降迹象;定位地下结构或孔穴、埋管、公共设施;提供样本供进一步检查及实验室测试。

图 2.2 现场勘探和取样

(Figure 2.2 Site exploration and sampling)

(4) Laboratory testing of samples: tests on disturbed and undisturbed samples submitted from the site team; tests on soils for classification, quality, permeability, shear

strength, compressibility, etc.; tests on rock cores and samples for strength and durability; test on constructional material, such as California bearing ratio; tests on groundwater; chemical and petrographic analyses.

(4) 样品的实验室测试：测试工地小组提交的扰动及未扰动样品；测试土壤的分类、质量、渗透性、抗剪强度、可压缩性等；测试岩心和样品的强度和耐久性；测试建筑材料，例如加利福尼亚承载比；测试地下水；进行化学和岩相分析。

(5) In situ testing (Figure 2.3): tests carried out on the site either prior to or during the construction process; ground tests such as shear vane, standard penetration, cone penetration, plate bearing, pressuremeter; structure loading texts, such as tests on piles, proof loading; displacement observations.

(5) 现场测试(图 2.3)：施工前或施工期间在工地进行的测试；地面测试，例如剪对切叶片、支柱打进、锥体贯穿、平板轴承、压力计；结构加载文本，例如桩系测试、验证加载测试；位移观测。

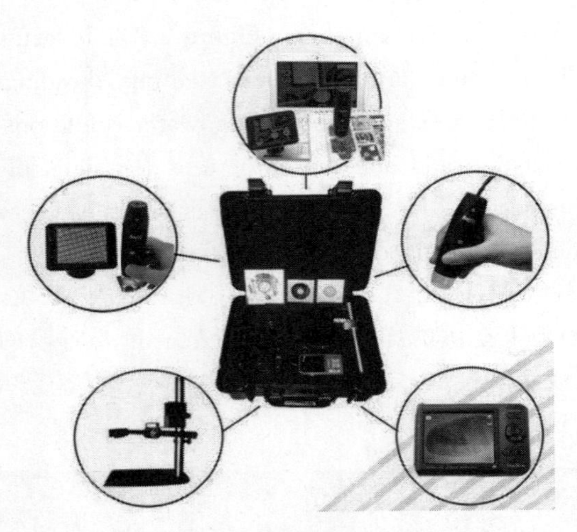

图 2.3　现场测试工具概观

(Figure 2.3　An overview of field test tools)

(6) Reporting results: details of geological study, including structures, stratigraphy and mapping; results of borings, etc., including log, references for samples and stratigraphy interpretations as requested; comments and recommendations relating to the design and construction of the proposed works; recommendations relating to further investigating or testing, and to ongoing-or post-completion monitoring.

(6) 成果报告：地质研究详情，包括结构、地层和测图；钻孔结果等，包括(勘探)日志、样品参考资料和地层解释；与拟议工程的设计及施工有关的意见及建议；与进一步的调查或测试、进行中或完成后的监测有关的建议。

2.1.2 场地调查报告(Site investigation reports)

A site investigation report is usually the culmination of the investigation, exploration and testing program, although intermediate reports may sometimes be required where long-term onongoing observation are involved. The report will be addressed to the client or whomever has commissioned the investigation; it may be purely factual or may contain advice and recommendations relating to design and construction, and sometimes suggestions relating to post-construction monitoring.

场地调查报告通常是调查、勘探和测试程序的归纳总结,如果涉及长期持续观察的,有时可以要求中期报告。该报告将由客户或委托进行调查的任何人来处置;它可以是纯事实的,也可包含有关设计和施工的忠告和建议,有时还包含有关施工后监测的建议。

Although individual reports vary according to the particular brief received and conditions encountered, a typical report will normally include the following:

虽然个别报告会根据收到的简报及碰到的状况而有所变化,但一份典型的报告正常情况下会包括以下内容:

(1) Introduction: a brief summary of the proposed works, the investigations carried out, the location of the site and significant names and dates.

(1) 引言:对所提建议工程的简要概括、所进行的调查、工地位置、重要名称和日期。

(2) Description of site: a general description of the site: its topography and main surface features; details of access; details of previous development or relevant history, details of existing works, underground openings, drainage, etc.; a map showing site location, adjoining land and borehole locations.

(2) 场地描述:普通描述包括:地形及主要地表特点;通道细节;先前的开发或有关史料的详情;现有工程、地下开口、排水系统等的详情;显示现场位置、毗邻土地和钻孔部位的地图。

(3) Geology of the site: commencing with a description of overall geology, related to the regional geology of the area; description of main soil and rock formations and structures, comments on the influences of geology on design and construction.

(3) 场址地质:开篇即阐明有关"区域"中的区域地质的概况;描述主要岩土的形成及结构,评述地质对设计和施工的影响。

(4) Soil conditions: a detailed account of the soil conditions encountered, related to the design and construction of the proposed works, description of all relevant layers, together with results of laboratory and *in situ* tests; details of groundwater and drainage conditions.

(4) 土壤状况:详细说明有关拟议工程的设计和建造所面临的土壤情况,说明所有有关的土层,以及实验室和现场测试的结果;详述地下水及排水条件。

(5) Construction materials: a detailed account of the nature, quantity, availability and significant properties of materials considered for construction purposes.

(5) 施工材料:为施工考虑的材料的自然性质、数量、可用性和重要性质的详细说明。

(6) Comments and recommendations: comments are necessary on the validity and reliability of the information being presented; where further work is required this should be mentioned; if the brief is also to make recommendations, these should include consideration of alternative methods of both design and construction.

(6) 意见和建议：有必要对递交中信息的有效性和可靠性提出意见；凡要求进一步工作的，均宜予以提及；若简报也提出建议时，宜包括对考虑设计和施工两方面的替代方法。

(7) Appendices: it is convenient to assemble most of the collected data into a series of appendices: borehole logs; laboratory test details and results; results of in-situ tests; geophysical survey records; references; relevant literature extracts.

(7) 附录：将收集到的大部分数据汇编成一系列附录较为便利：钻孔测井；实验室测试详情和结果；现场测试的结果；地球物理勘查记录；引用；相关文献摘录。

2.2 基础开挖(Excavation for Foundation)

Excavation procedure for foundation construction requires site clearance, setting out, excavation and safety measures based on depth of excavation.

基础施工的开挖程序要求清理场地、放线、开挖以及根据开挖深度所采取的安全措施。

2.2.1 地基开挖前场地清理(Site clearance before excavation for foundation)

Before the excavation for the proposed foundation is commenced, the site shall be cleared of vegetation, brushwood, stumps of trees etc. Roots of the trees shall be removed to at least 30cm below the foundation level (Figure 2.4). The pits formed due to roots of trees, old foundations etc. shall be filled up with soil and compacted.

在挖掘拟建地基开始前，应清除该场地的植被、灌木、树桩等。树根应移除得至少低于基础水平 30cm(图 2.4)。因树根、老地基等所形成的坑凹应以土填满压实。

图 2.4 地基开挖前场地清理

(Figure 2.4 Site clearance before excavation for foundation)

2.2.2 布置开挖地基布局(Setting out foundation layout for excavation)

For setting layout of foundation excavation, a benchmark shall be established at the site by a masonry pillar and connected to the nearest standard benchmark. Levels of the site should be taken at 5 to 10 millisecond depending on the terrain and the importance of the building.

基础开挖布置时,应在现场用砌石柱建立一处基准,并与就近的标准基准连接。根据地形和建筑物的重要性,以 5～10m 间距对场地进行分层。

The center lines of the walls are marked by stretching strings across wooden pegs driven at the ends. The center lines of the perpendicular walls are marked by setting out the right angle with steel tapes or preferably witha theodolite.

墙的中心线通过两端拧/系木栓拉伸绳线来标记。用钢带或更偏爱用经纬仪按直角放线来标出垂直墙的中心线。

The setting out of walls shall be facilitated by having a permanent row of pillars (not less than 25cm side) parallelly laid at a suitable distance beyond the periphery of the building so that they do not foul with the excavation. The pillars shall be located at the junctions of the cross walls and external wall and shall be bedded sufficiently deep so that they are not disturbed during excavation for foundation.

为方便墙壁的放线应在建筑物外围合适的距离平行放置一排永久性柱子(侧面不少于 25cm),以免因开挖物弄脏。柱子应位于横墙和外墙的连接处,并应埋置够深,使其在基础开挖期间不受干扰。

The center lines of the walls shall be extended and marked on the plastered tops of the pillars. The tops of the pillars may be kept at the same level, preferably the plinth level. In rectangular or square settings, the diagonals shall be checked to ensure accuracy of setting out (Figure 2.5).

墙的中心线应予以延长,并在柱体石灰抹顶上标明。柱顶可保持同一水平,最好是柱基水平。在矩形或方形设置中,应检查对角线,以确保放线的准确度(图 2.5)。

图 2.5 开挖地基布局放线

(Figure 2.5 Setting out foundation layout for excavation)

2.2.3 地基开挖程序(Excavation procedure for foundation)

For small buildings, excavation is carried out manually by means of pickaxes, crow bars, spades etc. In case of large buildings and deep excavation, mechanical earth cutting equipment can be used.

对于小型建筑物,采用鹤嘴锄、撬棍、铁锹等工具人工开挖。而对大型建筑物与深基坑则能用机械挖掘设备。

For hard soils when the depth of excavation is less than 1.5m, the sides of the trench do not need any external support. If the soil is loose or the excavation is deeper, some sort of shoring is required to support the sides from falling.

对于坚硬土质,开挖深度小于1.5m时,壕沟两侧无须任何外部支撑。如果土质松散或挖掘更深,就要求某种支撑物来支撑侧边,以免塌陷。

Planking and strutting can be intermittent or continuous depending on the nature of soil and the depth of excavation. In the case of intermittent or open planking and strutting the entire sides of trenches are not covered.

铺板和支柱能采用断续式或连续式,这取决于土壤的自然性质和挖掘深度。在间歇或露天铺板和支柱的情况下,沟槽两侧全部不予覆盖。

Vertical boards (known as poling boards) of size 250mm×40mm of there required length can be placed with gaps of about 50cm. These shall be kept apart by horizontal wailings of strong timber of section 100mm×100mm at a minimum spacing of 1.2m and strutted by a cross piece of 100mm×100mm square or 100mm diameter.

垂直板(称为杆板)的尺寸为所需长度250mm×40mm,能以约50cm的间隙来放置。这些杆板应以截面为100mm×100mm的强固木质横桁隔开,最小间距1.2m,并由一块100mm×100mm方形或直径100mm的横桁支撑。

2.2.4 用于松软土壤中开挖的露天铺板(Open planking for excavations in soft and loose soils)

If the soil is very soft and loose, the boards shall be placed horizontally against the sides of the excavation and supported by vertical waling boards which shall be strutted to similar timber pieces on the opposite side of the trench. Care has to be taken while withdrawing the timber members after completion of the foundation work so that there is no collapse of trench.

如果土壤很软很松,这些木板应水平放置在开挖的两侧,并由垂直的支撑板支撑,这些支撑板应与壕沟的另一侧类似的木片支撑。基础工程完成后,在收回木构件时要小心,以免沟槽坍塌。

2.2.5 开挖区排水作业(Dewatering of excavation)

Construction of foundation below the subsoil water level poses problems of water logging. It is therefore very often necessary to dewater the area of excavation(Figure 2.6).

地下水位以下的地基施工提出了水涝问题。因此，经常有必要对开挖区排水(图 2.6)。

图 2.6 开挖区排水作业

(Figure 2.6 Dewatering of excavation)

Several operations have to be carried out within the excavation, like laying bed concrete, laying of RCC raft slab and construction of masonry etc.

开挖期内不得不铺混凝土床、铺装混凝土筏板，以及砌筑等施工。

Therefore, work can be carried out more efficiently if the excavation area is kept dry.

因此，如果挖掘区保持干燥，则能更有效地实施工作。

2.2.6 降低开挖底部以下的水位(Reducing water level below the excavation bottom)

To keep the area of excavation dry, water table should be maintained at least 0.5m below the bottom of the excavation. There are several methods available for lowering the water table. Information obtained from site and soil investigation would be useful in deciding the most suitable and economical method of dewatering.

为保持挖掘区干燥，宜将地下水位维持在挖掘区底部以下至少 0.5m。有几种降低水位的现成方法。可从场地和土壤调查中得到的信息来决定最合适最经济的排水方法会很有用。

2.2.7 浅地基排水作业(Dewatering for shallow foundations)

For fairly dense soil and shallow excavations, the simplest method is to have drains along theedges of the excavation and collect water in sump sand remove it by bailing or pumping. This is the most economical method and is feasible of being executed with unskilled labor and very simple equipment.

对于相当密集的土壤和浅挖掘，最简单的方法是沿挖掘边缘设置排水沟，将水收集到集水池中,并通过舀水或抽水将其移除。这是最经济可行的方法,用不熟练劳动力,设备也非常简单。

2.2.8 大型挖掘和深地基的排水作业(Dewatering for large excavations and deep foundations)

Where large excavations such as for rafts are to be dewatered, wellpoint system can be employed. Wellpoint consists of a perforated pipe, 120cm long and 4cm in diameter with a valve to regulate flow and a screen to prevent entry of mud etc.

在必须排水处理的大型挖掘区,例如浮木筏基础,能采用井点系统。井点由一段长120cm、直径4cm的穿孔管,一个阀门来调节流量和一张滤筛阻止泥浆进入管等组成。

These wellpoints are installed along the periphery of the excavation at the required depth and spaced at about 1m. The exact spacing can be decided on the basis of the type of soil.

这些井点安装在开挖处的外围,按深度和要求间距约1m。恰当的间距能根据土壤的类型来决定。

Wellpoints are surrounded by sand gravel filter and have riser pipes of 5 to 7.5cm diameter. These pipes are connected to a header pipe which is attached to a high capacity suction pump. The groundwater is drawn out by the pumping action and is discharged away from the site of excavation.

井点有沙砾滤层,并有直径5～7.5cm的立管组。这些管组连接到一个头管,再接到一台高水量的吸泵。地下水由泵吸作用抽出,从挖掘场地排掉。

2.2.9 地基混凝土施工(图2.7)(Concreting of foundation in excavation, Figure 2.7)

图2.7 地基混凝土施工

(Figure 2.7 Concreting of foundation in excavation)

In the case of a masonry wall, the footing is generally of cement concrete mix of ratio 1∶4∶8 or 1∶5∶10 (cement∶sand∶coarse aggregate). The size of coarse aggregate is limited to 40mm. Lime concrete can also be used for this purpose.

在砌体墙情况，基脚一般用混合比1∶4∶8或1∶5∶10(水泥∶沙子∶粗骨料)的混凝土。粗骨料的尺寸限制在40mm以内。石灰混凝土也能用于此目的。

For important works, mixing of concrete should be done in a mechanical mixer. Concrete should be laid (not thrown) in layers not exceeding 15cm and well compacted.

对于重要作业，混凝土搅拌宜由机械搅拌机承担。混凝土宜分层铺设(而非抛洒)，层厚不超过15cm，并压实压紧。

The concrete should lie protected by moist gunny bags after about 1 or 2 hours of laying. Regular curing should be started after 24 hours and be continued for 10 days.

混凝土铺设约1～2h，宜铺上潮湿的麻袋加以保护，24h后开始常规养护，持续10天。

The masonry work over the bed concrete can be started after 3 days of laying the concrete but curing along with that of masonry shall be continued.

床上混凝土砌筑3天后即能开始，但应随砌体持续养护。

For RCC column footings and raft foundations, a levelling course of lean concrete of 75mm is laid in order to have an even and soil-free surface for placing the reinforcement.

对于碾压混凝土柱基脚和筏板基础，铺设了75mm的贫混凝土平整层，以便有一个平坦、无土的表面来放置钢筋。

2.3 专业词汇(Specialized Vocabulary)

foundation　n. 基础，地基
pit　n. 坑
geological　adj. 地质的
geology　n. 地 质学
topography　n. 地形，地势
stream flows　(河水)流量
groundwater　n. 地下水
sub-surface soil　地表下土
geophysical　adj. 地球物理(学)的
shear strength　抗剪强度，剪切强度
petrographic　adj. 岩相的
pressuremeter　n. 压力表
displacement　n. 位移
exploration　n. 勘探
drainage　n. 排水(系统)
structures　n. 结构
soil　n. 土(壤)

construction materials 施工材料
excavation n. 挖掘
benchmark n. 基准(点)
wall n. 墙(体)
pillar n. 柱(体)
pickaxe n. 鹤嘴锄
spade n. 铁锹
gravel n. 沙砾
concreting n. 混凝土作业

习题(Exercises)

1. Translate the following sentences into Chinese.

(1) A thorough and comprehensive site investigation is an essential preliminary to the design and construction of a civil engineering project.

(2) Excavation procedure for foundation construction requires site clearance, setting out, excavation and safety measures based on depth of excavation.

(3) For setting layout of foundation excavation, a benchmark shall be established at the site by a masonry pillar and connected to the nearest standard benchmark.

(4) The centre lines of the walls shall be extended and marked on the plastered tops of the pillars.

(5) Care has to he taken while withdrawing the timber members after completion of the foundation work so that there is no collapse of trench.

2. Translate the following sentences into English.

(1) 为建筑目的考虑的材料的性质、数量、可用性和重要特性的详细说明。

(2) 在挖掘拟建地基前,应清除该地点的植被、灌木、树桩等。

(3) 用钢带或用经纬仪画出直角来标出垂直墙的中心线。

(4) 对于小型建筑物,使用人工开挖,采用鹤嘴锄、撬棍、铁锹等。

(5) 精确的间距可以根据土壤的类型来确定。

地基工程

(Foundation Engineering)

3.1 浅地基(Shallow Foundation)

3.1.1 引言(Introduction)

It is the customary practice to regard a foundation as shallow if the depth of the foundation is less than or equal to the width of the foundation. The different types of footings that we normally come across are given in Figure 3.1. A foundation is an integral part of a structure. The stability of a structure depends upon the stability of the supporting soil. Two important factors that are to be considered are.

根据惯例,如果地基的深度小于等于地基的宽度,则视为浅地基。我们通常遇到的不同类型浅基脚如图 3.1 所示。地基是建筑物的组成部分之一。建筑物的稳定性取决于支撑土的稳定性。要考虑的两个重要因素是:

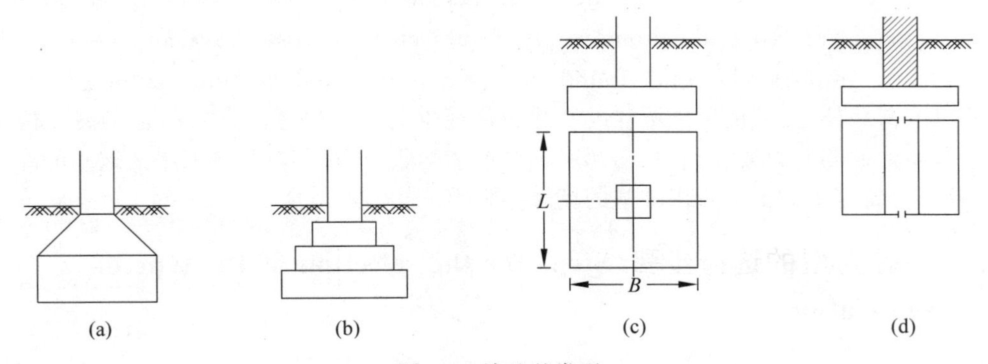

图 3.1　浅地基类型

(a) 素混凝土地基;(b) 阶梯式钢筋混凝土地基;(c) 钢筋混凝土矩形地基;(d) 钢筋混凝土墙地基

(Figure 3.1　Types of shallow foundations)

(a) plain concrete foundation; (b) stepped reinforced concrete foundation;

(c) reinforced concrete rectangular foundation; (d) reinforced concrete wall foundation

(1) The foundation must be stable against shear failure of the supporting soil.

(1) 地基必须是稳定的,以抵抗支承土的剪切失效。

(2) The foundation must not settle beyond a tolerable limit to avoid damage to the structure.

(2) 地基的沉降不能超过容许限度，以免结构发生破坏。

The other factors that require consideration are the location and depth of the foundation. In deciding the location and depth, one has to consider the erosions due to flowing water, underground defects such as root holes, cavities, unconsolidated fills, ground water level, and presence of expansive soils.

需要考虑的其他因素是地基的位置和深度。在决定其位置和深度时，必须考虑由流水引起的侵蚀、还要考虑地下缺陷如：根穴、孔洞、未固结的填充物、地下水位、膨胀土所产生的危害。

In selecting a type of foundation, one has to consider the functions of the structure and the load it has to carry, the subsurface condition of the soil, and the cost of the superstructure.

选择地基的类型时，必须考虑结构的功能及其承载的荷载、土壤的地下状况以及上部结构的成本。

Design loads also play an important part in the selection of the type of foundation. The various loads that are likely to be considered are (i) dead loads, (ii) live loads, (iii) wind and earthquake forces, (iv) lateral pressures exerted by the foundation earth on the embedded structural elements, and (v) the effects of dynamic loads.

设计荷载在选择基础的类型中也起重要作用。可能考虑的各种荷载有①恒载，②活载，③风和地震荷载，④地基土对嵌入结构单元施加的侧压力，⑤动荷载。

In addition to the above loads, the loads that are due to the subsoil conditions are also required to be considered. They are (i) lateral or uplift forces on the foundation elements due to high water table, (ii) swelling pressures on the foundations in expansive soils, (iii) heave pressures on foundations in areas subjected to frost heave and (iv) negative frictional drag on piles where pile foundations are used in highly compressible soils.

除以上荷载外，由于地下土壤状况引起的荷载也要求加以考虑。这些是①高水位引起的对地基要件的侧向力或上托力，②膨胀土对地基的胀压力，③对易受冻胀区地基的隆起压力，以及④对高可压缩土中所用桩基的负摩擦阻力。

3.1.2 地基类型的选择步骤(Steps for the selection of the type of foundation)

In choosing the type of foundation, the design engineer must perform five successive steps.

选择地基类型中，设计工程师必须连续实施5步。

(1) Obtain the required information concerning the nature of the superstructure and the loads to be transmitted to the foundation.

(1) 获取有关上层结构性质和传递给基础的荷载的所需信息。

(2) Obtain the subsurface soil conditions.

(2) 获取地下土壤条件。

(3) Explore the possibility of constructing any one of the types of foundation under the existing conditions by taking into account (i) the bearing capacity of the soil to carry the required load, and (ii) the adverse effects on the structure due to differentia settlements. Eliminate in this way, the unsuitable types.

(3) 通过考虑①土壤承载受所要求荷载的承受能力和②沉降差异对结构的不利影响，探讨在现有状况下建造任何一种类型地基的可能性。用这种方式剔除不合适的类型。

(4) Once one or two types of foundation are selected on the basis of preliminary studies. make more detailed studies. These studies may require more accurate determination of loads, subsurface conditions and footing sizes. It may also be necessary to make more refined estimates of settlement in order to predict the behavior of the structure.

(4) 在初步研讨的基础上一旦选出了一两种类型，就需要进行更详细的研究。这些研究可能需要更准确地确定荷载、地下条件和基础尺寸。为了预测结构的性能，也有必要对沉降做出更精确的估计。

(5) Estimate the cost of each of the promising types of foundation, and choose the type that represents the most acceptable compromise between performance and cost.

(5) 估计看好的每一处地基类型的成本，并选出在性能和成本之间体现最易接受的折中类型。

3.1.3 浅地基(Shallow foundations)

A shallow foundation generally isdefined as a foundation that bears at a depth less than about two times its width. There is a wide variety of shallow foundations. The most commonly used ones are isolated spread footings, continuous strip footings, and mat foundations.

浅地基一般定义为深度小于其宽度两倍的地基。浅基础类型繁多，最常用的有独立地基、连续条形地基和垫层地基。

Many shallow foundations are placed on reinforced concrete pads or mats, with the bottom of the foundation only a few feet below the ground surface. The engineer will select the relatively inexpensive shallow foundation for support of the applied loads if analyses show that the near-surface soils can sustain the loads with an appropriate factory of safety and with acceptable short-term and long-term movement. A shallow excavation can be made by earth-moving equipment, and many soils allow vertical cuts so that formwork is unnecessary. Construction in progress of a shallow foundation is shown in Figure 3.2. The steel seen in the figure may be dictated by the building code controlling construction in the local area.

许多浅地基置于钢筋混凝土垫块上，地基底部距地面仅几英尺。当分析表明，近地表土以适当的安全因数有可接受的短期和长期运动，能承受各种荷载时，工程师将选择相对便宜的浅地基来支撑所施加的荷载。挖土设备能浅挖作业，许多土壤允许垂直切挖，因此无须模板。浅基础施工进展如图 3.2 所示。图中所见钢材可以选控制当地施工的建筑规范所规定的。

图 3.2 进展中的浅地基施工

(Figure 3.2 Construction of a shallow foundation in progress)

Shallow foundations of moderate size will be so stiff that bending will not cause much internal deformation, and such foundations are considered rigid in analyses. The distribution of stress for eccentric loading is shown in Figure 3.3(a), and bearing-capacity equations can be used to show that the bearing stress at failure, q_{ult} provides an appropriate factor of safety with respect to q_{max}.

中等大小的浅地基会很坚硬,弯曲也不至于引起很大内部变形,在分析中认为这样的基础是刚性的。偏心加载时应力分布如图 3.3(a)所示,承载能力方程能用来表示失效时的承载应力,q_{ult}对于 q_{max}提供了适当的安全因数。

Shallow foundations can also be designed to support horizontal loads, as shown in Figure 3.3(b). Passive pressure on the resisting face of the footing and on the surface of a key, along with horizontal resistance along the base of the footing, can be designed to resist the horizontal load. Active pressure would occur on faces moving away from the soil, but these may be ignored as being too small to make any difference in the solution.

浅地基也可能用来支撑水平荷载,如图 3.3(b)所示。在基脚的抵挡面和拱顶石表面上的被动压力,连同沿基脚底部的水平阻力,能设计来抵抗水平荷载。主动压力会发生在离开土壤一定距离的面上,但这些可予忽略,因其太小,对解决方案没有任何影响。

Usually, shallow foundations are less expensive than deep foundations, but designs become more complicated as the foundation becomes larger in plan. Significant stress for a mat or larger shallow foundation reaches deeper soils, and the computation of deformation becomes more complicated than for the foundation of moderate size. Not only will the vertical movement be larger than for a footing, but the deformation of the mat must be considered as well as the deformation of the supporting soil.

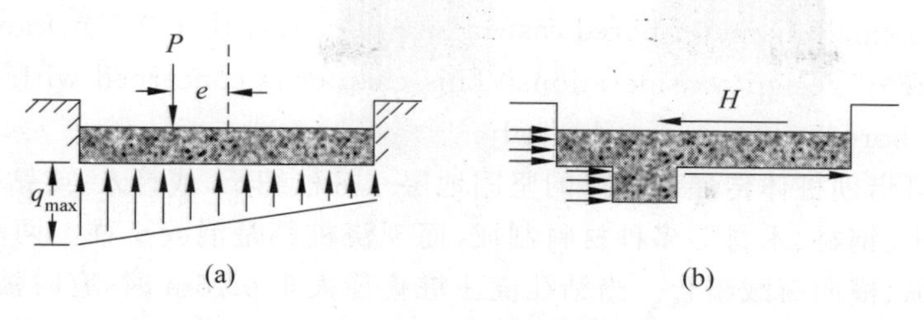

图 3.3 浅地基立面图

(a) 偏心竖向荷载作用下地基的承受应力;(b) 横向荷载作用下浅地基的应力分布

(Figure 3.3 Elevation views of shallow foundations)

(a) Bearing stress for a foundation under eccentric vertical loading; (b) Stress distribution for a shallow foundation under horizontal loading

浅地基成本通常比深地基的低,但是随着基础在规划上变大,设计就更加复杂。对于垫层或大型浅基础,有效应力达到更深的土体,其变形计算比中等大小基础的更为复杂。不仅是垂直移动比基脚的大,而且必须考虑垫层变形以及支撑土的变形。

The principal problem with shallow foundations under light to moderate loading concerning expansive clay is widespread and can be devastating to homeowners. Engineers must be especially diligent in identifying expansive clay at a building site and taking appropriate actions if such soil is present.

轻、中型荷载作用下浅基础所涉及膨胀土问题很普遍,对房主来说能造成毁灭性的影响。工程师们必须特别尽职地在建筑工地辨识膨胀黏土,并在出现这种土壤时采取行动。

3.2 深基础:桩基础(Deep Foundation:Pile Foundation)

3.2.1 引言(Introduction)

Shallow foundations are normally used where the soil close to the ground surface and up to the zone of significant stress, and possess sufficient bearing strength to carry the superstructure load without causing distress to the superstructure due to settlement. However, where the top soil is either loose or soft or of a swelling type the load from the structure has to be transferred to deeper firm strata.

浅地基正常情况下用于从靠近地面之处,直至应力显著的地带,并具有足够承载强度来背负上层结构荷载,而不致因沉降而对上层结构造成损害。然而,在顶层土或松或软或属膨胀型之处,来自此结构的荷载必须迁移到更深的坚实地层。

The structural loads may be transferred to deeper firm strata by means of piles. Piles are long slender columns either driven, bored or cast-in-situ. Driven piles are made ofa variety of materials such as concrete, steel, timber etc., whereas cast-in-situ piles are concrete piles. They may be subjected to vertical or lateral loads or a combination of vertical and lateral loads. If the diameter of a bored-cast-in-situ pile is greater than about 0.75m, it is sometimes called a drilled pier, drilled caisson or drilled shaft. The distinction made between a small diameter bored cast-in-situ pile (less than 0.75m) and a larger one is just for the sake of design considerations. This chapter is concerned with driven piles and small diameter bored cast-in-situ piles only.

结构荷载可借助桩体传递到更深的坚固地层。桩柱细长，或打入，或钻孔，或现场浇筑。打入桩由混凝土、钢材、木材等多种材料制成，而现浇桩都是混凝土柱。两者均可承受纵向或横向荷载或纵、横向荷载组合。当钻孔浇注桩直径大于0.75m时，有时被称为钻孔式墩、钻孔式沉箱或钻孔式井。小直径钻孔浇注桩(小于0.75m)和大直径的两者的区别只是出于设计考虑。本章仅涉及打入桩和小直径钻孔浇注桩。

3.2.2 桩的分类(Classification of piles)

Piles may be classified as long or short in accordance with the L/d ratio of the pile (where $L=$ length, d =diameter of pile). A short pile behaves as a rigid body and rotates as a unit under lateral loads. The load transferred to the tip of the pile bears a significant proportion of the total vertical load on the top. In the case of a long pile, the length beyond a particular depth loses its significance under lateral loads, but when subjected to vertical load. the frictional load on the sides of the pile bears a significant part to the total load.

桩按长径比L/d(其中：$L=$桩长，$d=$桩径)可分为长桩和短桩两类。短桩为刚体，在侧向荷载作用下整体转动。转移到桩顶的荷载在桩顶总垂直荷载中占很大比例。在长桩的情况下，超过某一特定深度的那段在横侧向荷载作用下失去作用，但当承受垂直荷载时会直接失去作用。桩侧摩擦荷载占桩总荷载的很大一部分。

Piles may further be classified as vertical piles or inclined piles, Vertical piles are normally used to carry mainly vertical loads and very little lateral load. When piles are inclined at an angle to the vertical, they are called batter piles or raker piles. Batter piles are quite effective for taking lateral loads, but when used in groups, they also can take vertical loads.

桩柱可进而分为垂直桩或斜桩两类，垂直桩通常主要承受垂直荷载，很少承受侧向荷载。当桩与垂直方向成一定角度倾斜时，称为倾斜桩或耙桩。倾斜桩对承受横向荷载非常有效，但在成组使用时，也能承受垂直荷载。

Piles may be classified according to their composition as:

桩可按其构成材料分类为：

(1) Timber piles.

(1) 木桩。

(2) Concrete piles.

(2) 混凝土桩。

(3) Steel piles.

(3) 钢桩。

Timber Piles: Timber piles are made of tree trunks with the branches trimmed off. Such piles shall be of sound quality and free of defects. The length of the pile may be 15m or more. If greater lengths are required, they may be spliced. The diameter of the piles at the butt end may vary from 30 to 40cm. The diameter at the tip end should not be less than 15cm.

木桩：由修剪掉枝杈的树干制成。此类桩应材质优良，不带瑕疵。桩长度可达15m或更长。要求更长时，可予以拼接。对接端桩直径在30～40cm变化。尖头端直径不宜小于15cm。

Piles entirely submerged in water last long without decay provided marine borers are not present. When a pile is subjected to alternate wetting and drying the useful life is relatively short unless treated with a wood preservative, usually creosote at 250kg/m^3 for piles in fresh water and 350kg/m^3 in sea water.

假如没有海洋钻蛀虫，桩完全浸入水中就可长时间不腐。桩体处于交替干湿的状态时，可用寿命相对缩短，除非用防腐剂进行处理，淡水中通常用250kg/m^{3}的杂酚油；海水中通常使用350kg/m^{3}的杂酚油。

After being driven to final depth, all pile heads, treated or untreated, should be sawed square to sound undamaged wood to receive the pile cap. But before concrete for the pile cap is poured, the head of the treated piles should be protected by a zinc coat, lead paint or by wrapping the pile heads with fabric upon which hot pitch is applied.

打入至最终深度后，所有桩头，经处理的或未经处理的，均宜锯成方形，锯至木质完好无损，以便套上桩帽。但在桩帽浇注混凝土前，处理过的桩头宜予保护，涂上锌层、铅漆料或用织物包裹，并敷上热沥青。

Driving of timber piles usually results in the crushing of the fibers on the head (or brooming) which can be somewhat controlled by using a driving cap, or ring around the butt.

木桩打入作业通常导致桩头纤维破碎或蓬裂，这可以用抗打帽或在对接处加箍来进行某种程度的控制。

The usual maximum design load per pile docs not exceed 250kN. Timber piles are usually less expensive in places where timber is plentiful.

单桩最大设计荷载通常不超过250kN。木材充裕之处，木桩往往便宜。

Concrete piles: concrete piles are either precast or cast *in situ* piles. Precast concrete piles are cast and cured in a casting yard and then transported to the site of work for driving. If the work is of a very big nature, they may be cast at the site also.

混凝土桩：混凝土桩分为预制桩和浇制桩。预制混凝土桩在浇筑场浇制并养护，然后运至工地进行打桩。如果这项作业工作量很大，也可在工地浇制。

Precast piles may be made of uniform sections with pointed tips. Tapered piles may be manufactured when greater bearing resistance is required. Normally piles of square or octagonal sections are manufactured since these shapes are easy to cast in horizontal position. Necessary reinforcement is provided to take care of handling stresses. Piles may also be prestressed. Maximum load on a prestressed concrete pile is approximately 2000kN and on precast piles 1000kN. The optimum load range is 400 to 600kN.

预制桩可由带尖端的均匀截面制成。当要求更大的承载力时，可制造锥形桩。正常情况，截面制成方形或八角形，因为这些形状容易在水平位置制造。提供必要的加固以兼顾处理应力。桩也可以作预应力处理。预应力混凝土桩的最大荷载约为 2000kN，预制桩最大荷载约 1000kN。最佳荷载范围 400～600kN。

Steel Piles. Steel piles are usually rolled H shapes or pipe piles. H-piles are proportioned to withstand large impact stresses during hard driving. Pipe piles are either welded or seamless steel pipes which may be driven either open-end or closed-end. Pipe piles are often filled with concrete after driving, although in some cases this is not necessary, The optimum load range on steel piles is 400 to 1200kN.

钢桩：通常轧制工字钢或管桩。工字桩身按比例分布，以抵挡硬打桩时的大冲击应力。管桩有焊接钢管和无缝钢管两种，可开端或闭端打入。管桩在打入后往往以混凝土填充，（虽然在某些情况下这并不必要），钢桩的最佳荷载范围是 400～1200kN。

3.3 地基改善(Ground Improvement)

3.3.1 介绍(Introduction)

The design of each foundation is a unique challenge for engineers. Designs must take into account the nature and cost of the structure, geology and terrain, quality of subsurface investigation, type of protocol with the owner, loadings over the life of the structure, effects of the proposed construction on buildings near the construction site, sensitivity of the proposed structure to total and differential settlement, requisite building codes, potential environmental effects due to the proposed construction including excessive noise, adequacy of analysis tools available for design, loss of money or loss of life as a result of a malfunction, availability of materials for foundations, and competent contractors in the area. A considerable effort will normally be necessary to gather and analyze the large amount of relevant data.

每一基础的设计对工程师们来说都是一次独特的挑战。设计必须考虑：结构的自然性质和成本，地质和地形，地表下调查的质量，与业主协议的类型，结构寿命期的荷载，拟议施工对施工现场附近建筑物的效应，拟建结构对总沉降和差别沉降的敏感度，必备建筑法规，拟议施工造成的潜在环境效应（包括过度噪声），可用于设计的分析工具的适当性，功能失常造成的财产或生命损失、地基材质可用性，以及该地的胜任承包商。正常情况下，有必要付出极大努力去收集和分析大量有关数据。

To create a foundation for good performance, the engineer must carefully address the topics and other related factors in the above list. Failure will occur if the foundation is not constructed in time as planned, if hazardous settlement occurs at any time after construction, adjacent buildings are damaged, or design waste.

为了建造性能良好的基础，工程师必须仔细处置上述清单中的议题和其他有关因素。如果地基未按计划及时施工，如果施工后不定什么时间出现有害沉降，相邻建筑物受损或设计浪费，均会导致失败。

The topic of designing the wasteful foundation is interesting. Engineers can spend more resources than necessary without peer review. The engineer increased the size of a foundation because the extra material gave a measure of "pillow comfort". An engineer from a company was asked to describe the technique used to design a pile for an important job and replied, "we use an approximate method and always add some piles to the design." No one had ever seen them, but they were failures.

设计基础浪费的话题很有趣。在未经同行评审情况下,工程师们耗费的资源比必要的多。工程师增大了地基的尺寸,因为额外的材料给出一个"枕头安慰"的手段。某公司一位工程师应邀描述用于为重要工作设计一根桩的技术,他答道:"我们采用了近似法,是设计中总追加一些桩。""从来没人见过这些基础,却都是失效桩。"

Casa Grande wrote down calculated risks in earthwork and foundation works, and listed several examples of risk projects used in the design and construction of major projects. There is always a risk in geotechnical engineering because good information about these factors is not available and sometimes due to abnormal behaviour in the soil, it is impossible to predict despite due diligence. Prudent engineers use available tools and techniques and keep abreast of technological advances to minimize risk.

卡萨·格兰德在土方工程和基础作业中写下了算出的风险,并列出了重大工程设计施工中风险项目的实例。岩土工程中总带某种风险,因为无法获得有关上述因素的足够信息,而且有时土壤的性态异常,尽管尽职尽责,仍无法预测得到。谨慎的工程师使用可用的工具和技术,并与技术进步保持同步,将风险降至最低。

Peck presented a keynote address at a conference where the emphasis was on improved analytical procedures. He gave five sources of error with regard to the bearing capacity and settlement of foundations: (1) the assumed loading may be incorrect; (2) the soil conditions used in the design may differ from the actual soil conditions; (3) the theory used in the design may be wrong or may not apply; (4) the supporting structure may be more or less tolerant to differential movement; and (5) defects may occur during construction. The technical literature is replete with examples of foundations that have failed, many for reasons noted by Peck. The following sections present brief discussions of some examples, emphasizing the care to be taken by the engineer in planning, designing, and specifying methods of construction of foundations.

派克在强调改进分析程序的一次会议上发表了主旨演讲。关于地基的承载能力和沉降,他给出了误差的5种来源:①假定的荷载可以不正确;②设计中所用土壤状况可异于实际;③设计中采用的理论可以有错误或不适用;④支撑结构可或多或少地容许差异运动;以及⑤施工期间可出现缺陷。技术文献中充斥已失败基础的实例,其中许多根源是派克提到的。以下几节简要讨论了一些实例,强调了工程师在规划、设计和规定基础施工方法时应注意的事项。

Lacy and Moskowitz made the following recommendations for deep foundations: detailed understanding of underground conditions; select a qualified contractor; pay attention to the compilation of construction specifications; full-time inspection by knowledgeable personnel with the necessary authorities; the adjacent structures are monitored during construction.

Lacy 和 Moskowitz 就深地基提出了以下建议：详细了解地下状况；选择有资质的承包商；注意编制施工规范；由知识渊博兼具必要权威的人士专职进行检查；施工期间监测邻近结构。

3.3.2 总沉降和差异沉降(Total and differential settlement)

The total settlement of a structure is the maximum amount the structure has settled with respect to its original position. The Palace of Fine Arts in Mexico has settled several feet but still remains usable because the differential settlement is tolerable. Differential settlement causes distortions in a structure, possible cracks in brittle materials, and discomfort to the occupants. The masonry building materials used in constructing the Palace of Fine Arts can tolerate the inevitable differential settlement, but some discomfort is inevitable. One can feel a definite tilting while sitting in a seat and watching a performance.

建筑物的总沉降是相对于其本原位置的最大沉降。墨西哥美术宫已沉降了好几英尺，但仍可继续使用，因差异沉降容许。差异沉降导致结构扭曲，脆性材料可能破裂，并对住户造成困扰。建造美术宫所用砌筑材料，能容许无法避免的差异沉降，造成一些不适在所难免。坐在席位上观看表演时，能感到一种明显的倾斜。

Subsequent sections of the foundation settlement calculation indicate that the maximum settlement should occur at the minimum values below the center and edge of the foundation. Some suggest that the difference between the maximum and minimum, the differential, should be part of the maximum. However, in all but exceptional cases, the engineer shall design a design with a moderate total settlement and negligible differential settlement.

地基沉降计算的后续几节指明，最大沉降会发生在地基的中心和边缘以下的最小值处。一些建议认为，最大值和最小值之间的差值，即差异，宜归入该最大值的组成部分。然而，除特殊情况外，所有情况下工程师都应设计得总沉降适中，差异沉降可忽略不计。

Settlement as a result of subsidence can be important in the design of foundations. Subsidence occurs because water or perhaps oil is removed from the underlying formations and causes an increase in effective stress, triggering settlement that can amount to several feet. Subsidence has occurred in many cities because of dewatering. Houston, Texas, is an example. The movement of the surface usually is relatively uniform and foundation problems are minimal, but faults can occur suddenly and can pass through the foundation of a home, with serious and perhaps disastrous consequences. A less common cause of subsidence is the yielding of supports for tunnels used in mines that have been abandoned in some regions of the United States.

沉降作为下沉的结果，在地基设计中很重要。下沉的起因是水，也许石油，从下伏地层中移开并导致有效应力增加，能引发多达几英尺的沉降。由于排水作业，许多城市都发生了沉降。德克萨斯州休斯敦就是一例。地表移动通常相对均匀,基础下沉问题是最小的,但断层能突然发生,并能贯穿屋基,带来严重甚至灾难性后果。另一种不大常见的下沉原因是,在美国一些地区废弃矿井用过的隧道支架垮塌。

Subsidence areas may become more frequent in the future as the demand for water increases and may lead to more pumping from aquifers. The engineer must understand the area where subsidence is occurring or is expected to occur.

随着对水的需求增加,以及可导致从含水层的抽水更多,未来下沉区会更为常见。工程师必须了解正在或预期发生下沉的区域。

3.4 专业词汇(Specialized Vocabulary)

foundation n. 基础,地基
shallow foundation 浅基础
soil n. 泥;土壤
structure n. 结构;构造;建筑物
ground water level 地层水位
unconsolidated fill 膨胀土
dead load 静荷载
live load 活荷载
wind and earthquake forces 风和地震力
dynamic loads 动力荷载
plain concrete foundation 素混凝土基础
stepped reinforced concrete foundation 阶梯式钢筋混凝土基础
reinforced concrete rectangular foundation 钢筋混凝土矩形基础
reinforced concrete wall foundation 钢筋混凝土墙基础
isolated spread footing 独立基脚
strip footing 条形地基
mat foundation 筏形基础
stratum(pl. strata) n. 层;地层;阶层
pile n. 桩
vertical pile 垂直桩
inclined pile 斜桩
timber pile 木桩
concrete pile 混凝土桩
steel pile 钢桩
settlement n. 沉降
brittle material 脆性材料
tunnel n. 隧道;地下道;洞群
subsidence n. 下沉;沉淀;陷没

aquifer n. 含水层

engineer n. 工程师

习题(Exercises)

1. Translate the following sentences into Chinese.

(1) It is the customary practice to regard a foundation as shallow if the depth of the foundation is less than or equal to the width of the foundation.

(2) The foundation must be stable against shear failure of the supporting soil.

(3) The foundation must not settle beyond a tolerable limit to avoid damage to the structure.

(4) The structural loads may be transferred to deeper firm strata by means of piles.

(5) The design of each foundation is a unique challenge for engineers.

2. Translate the following sentences into English.

(1) 获得有关上层结构的性质和传送至地基的荷载的所需资料。

(2) 根据桩的长径比L/d(其中：L=桩长,d=桩径),桩可分为长桩和短桩。

(3) 为了创建性能良好的基础,工程师必须仔细处理上述清单中的主题和其他相关因素。

(4) 建筑物的总沉降是相对于其本原位置的最大沉降。

隧道工程 (Tunnel Engineering)

Tunnels are passages which are built on rock masses, soil masses or under water, with entrances and exits at both ends for vehicles, pedestrians, water and pipelines to pass (Figure 4.1). Including railways, roads, underwater (sea) tunnels and various hydraulic tunnels in transportation.

隧道是修筑在岩体、土体或水下，两端有出入口，供车辆、行人、水流及管线通过的通道(图 4.1)。包括铁道、道路、水(海)底的隧道和各种水工运输隧道。

图 4.1 隧道示意图

(Figure 4.1 Schematic diagram of tunnel)

4.1 隧道结构(Tunnel Structure)

The structure of the tunnel includes two parts: the main building and auxiliary equipment. The main building is composed of a tunnel body and a tunnel door. The ancillary equipment includes shelters, fire-fighting facilities, emergency communication and water-proof and drainage facilities. The long tunnel also has special ventilation and lighting equipments.

隧道结构包括主体建筑和附属设备两部分。主体建筑由洞身和洞门组成。附属设备包括避车洞，消防设施，以及应急通信和防排水设施。长大隧道还有特殊的通风照明设备。

4.1.1 洞身衬砌(Tunnel lining)

Type of tunnel linings: (1) Straight wall lining: Straight wall lining is usually used when the vertical surrounding rock pressure of the rock formation is the main calculation load and the horizontal surrounding rock pressure is small; (2) Curved wall lining: Generally, the horizontal pressure is relatively high in the surrounding rock below W level. In order to resist the relatively large horizontal pressure, the side wall is also made into a curved shape; (3) Bias pressure lining: when the mountain slope is steeper than 1 : 2.5, the mountain cover outside the line is thin, or the bias pressure is caused by geological structure, the lining is used to withstand the pressure of such asymmetric surrounding rock; (4) Bell-mouth tunnel lining: Sometimes to bypass difficult terrain or avoid complex geological sections, to reduce engineering in mountain double-track tunnels, this method is applied; (5) Rectangular section lining: when the immersed pipe method is used for construction, the section can be in rectangular form, when the open-cut method is used for construction, especially when building multi-lane tunnels, rectangular is widely used in the section form; (6) Shotanchor lining, shotanchor lining and composite lining: In order to rationalize the stress state of the shotcrete structure, it is required to excavate with smooth blasting to make the surrounding of the cave smooth and accurate, and reduce over-excavation. Then spray concrete at an appropriate time, that is, spray concrete lining. According to the actual situation, if it is necessary to install the anchor rod, install the anchor rod first, and then spray the concrete, that is the spray anchor lining. One or several combinations of shotcrete, anchor rods or steel arch brackets are used as the primary support to strengthen the surrounding rock, maintain the stability of the surrounding rock and prevent harmful loosening. After the deformation of the primary support is basically stabilized, the cast-in-place concrete secondary lining is performed, that is a composite lining.

洞身衬砌的类型：①直墙式材砌：通常用在岩石地层中主要计算荷载为垂直围岩压力，而水平围岩压力不大的情况；②曲墙式衬砌：通常在W级以下围岩中，水平压力相对较大，为抵抗较大的水平压力，把边墙也做成曲面状；③偏压衬砌：当山坡陡度大于1∶2.5，线路外侧山体覆盖较薄，或由于地质构造造成偏压时，采用此衬砌以抵抗这一不对称围岩压力；④喇叭口隧道衬砌：在山区双线隧道，有时为绕过困难地形或避开复杂地质地段，减少工程量而选用此类设计；⑤矩形断面衬砌：用沉管法施工时，其断面能用矩形形式，用明挖法施工时，尤其在修筑多车道隧道时，其断面广泛采用矩形；⑥喷混凝土衬砌、喷锚衬砌及复合式衬砌：为了使喷射混凝土结构的应力状态合理，要求采用光面爆破开挖，使洞室周边平顺光滑，成型准确，减少超挖。然后在适当时机喷混凝土，此即喷混凝土衬砌。按照实际情况，有必要安装锚杆时先装设锚杆，再喷混凝土，此即喷锚衬砌。以喷混凝土、锚杆或钢拱支架中的一种或几种组合作为初次支护对围岩加固，维护围岩稳定并防止有害松动。待初次支护的变形基本稳定后，进行现浇混凝土二次衬砌，此即复合式衬砌。

4.1.2 支护结构(Supporting structure)

(1) Integral lining: a traditional lining structure;

(2) Composite lining: not only can give full play to the advantages of bolting and shotcrete support, but also can play a reliable role in the permanent support of the secondary lining;

(3) Shotcrete lining: As the permanent lining of the tunnel, the shotcrete support is generally considered to be used in the surrounding rock of Grade Ⅲ and above.

(1) 一体式衬砌:一种传统衬砌结构;

(2) 复合式衬砌:既能充分发挥锚喷支护的优势,又能发挥二次衬砌永久支护的可靠性;

(3) 喷混凝土式衬砌:作为隧道的永久衬砌,一般考虑是在Ⅲ级及以上围岩中采用。

4.1.3 洞门(Tunnel gate)

Tunnel portal (short for tunnel face, also generally refers to tunnel entrance section and open-cut-tunnel portal) is the exposed part at both ends of the tunnel. It is a retaining structure with masonry at the tunnel entrance to protect the hole, discharge water and decorate the building (Figure 4.2). It connects the lining and the road cut, is the main component of the entire tunnel structure, and is also a sign of the entrance and exit of the tunnel. Types of portals include ring frame, end-wall portal, wing wall portal, pillar portal, step portal, oblique portal, cut bamboo portal, etc.

洞门(隧道门的简称,也泛指隧道门及明洞门),是隧道两端的外露部分。是在隧道入洞口用圬工砌筑用以保护洞口、排放集水并对建筑物加以装饰的支挡结构(图 4.2)。洞门将衬砌和路堑相联系,是整个隧道结构的主要构件,也是隧道进出口标志。洞门类型包括洞口环框和端墙式、翼墙式、柱式、台阶式、斜交、削竹式等。

图 4.2 隧道(桥式)洞门

(Figure 4.2 Tunnel portal)

4.1.4 竖井与斜井(Shaft and inclined shaft)

Generally, tunnel excavation is carried out from the two openings or from one of the openings. However, due to construction period, economy, construction, topography, environment and other conditions, long and large tunnels must be divided into several engineering sections for construction. In most cases, working tunnels are required. The working tunnels are divided into horizontal holes, inclined shafts, vertical shafts and parallel pilot pits according to their slope.

一般来说,隧道开挖是从两处开口或其中一处开始进行。然而对大长隧道,因工期、经济、施工、地形、环境以及其他条件限制时,必须分成几个工程区段来施工,多数情况下要求设工作坑道。工作坑道按坡度可分为横洞、斜井、竖井和平行导坑。

4.2 隧道围岩分类和围岩压力(Tunnel Surrounding Rock Classification and Surrounding Rock Pressure)

4.2.1 围岩分类(Surrounding rock classification)

1. 隧道围岩分类方法(The method of tunnel surrounding rock classification)

①Classification method based on rock strength as a single lithology index; ②Classification method represented by rock mass structure and lithology characteristics; ③Classification method related to geological survey methods; ④Combination classification method of multiple factors; ⑤Classification method represented by engineering objects.

①按岩石强度单一岩性指标的分类法;②以岩体构造和岩性特征为代表的分类法;③与地质勘察手段关联的分类法;④按多种因素的组合的分类法;⑤以工程对象为代表的分类法。

2. 我国公路隧道围岩分类(Surrounding rock classification of highway tunnels in China)

The comprehensive evaluation method of tunnel surrounding rock classification in our country should adopt two-step classification, and proceed in the following order: ①According to the qualitative characteristics of the two basic factors of rock hardness and rock mass integrity, and the quantitative basic quality index (BQ) of rock mass, a comprehensive preliminary classification is carried out. ② When detailed grading of surrounding rock, the influence of correction factors should be considered on the basis of the basic quality classification of rock mass, and the BQ value of rock mass should be revised. ③ According to the revised BQ of the rock mass, judge the qualitative characteristics of the rock mass comprehensively to determine the detailed classification of the surrounding rock.

我国隧道围岩分类的综合评价方法宜采用两步分级,并按以下的顺序进行:①按岩石的硬度和岩体一体程度两个基本因素的定性特征和定量的岩体基本性质指标(BQ),进行综合性初步分级。②对围岩进行详细分级时,宜在岩体基本质量分类基础上考虑修正因素的影响,并修正岩体 BQ 值。③按修正后的岩体 BQ ,综合岩体的定性特征综合评判,确定围岩的详细分类。

4.2.2 围岩压力的产生(The generation of surrounding rock pressure)

The generation of surrounding rock pressure is an important mechanical feature of tunnel engineering. The tunnel is built in surrounding rock with a certain stress history and stress field. Therefore, the state of the initial stress field of the surrounding rock greatly affects all mechanical phenomena that occur in it, which is very different from ground engineering. We need to study the stress state of surrounding rock before and after tunnel excavation, which is of great significance for guiding our tunnel design and construction.

围岩压力产生是隧道工程的一个重要的力学特点。隧道是修筑在具有某种应力史和应力场的围岩中。因此,围岩初始应力场状态对围岩中产生的力学现象有极大的影响作用,这与地面工程大不相同。需要研究隧道开挖前后围岩的应力状态,这对指导隧道设计与施工有着重要意义。

4.2.3 围岩的初始地应力场(The initial stress field of surrounding rock)

The initial stress field generally refers to the initial static stress field of the rock mass before tunnel excavation. Its formation is closely related to the structure of the rock mass, the nature of the burial conditions, and the history of tectonic movement. It exists objectively before the tunnel is excavated. To build a tunnel in this stress field, it is necessary to understand its state and its influence. The initial stress state of the rock mass is different from the additional stress state caused by construction. It has an extremely important influence on the stress distribution, deformation and failure of the surrounding rock after the excavation of the tunnel. It can be said that without understanding the initial stress state of the rock mass, it is impossible to make a correct evaluation of a series of mechanical processes and phenomena after tunnel excavation. The initial stress state of the rock mass is generally affected by two types of factors: the first type of factors are gravity and temperature, the physical and mechanical properties of the rock mass, the structure of the rock mass, the topography and other regular factors; the second type of factors include temporary or local factors such as crustal movement, groundwater activities, and long-term human activities.

初始应力场通常是指隧道开挖前岩体的初始静应力场。其形成与岩体结构、埋藏条件的性质以及构造运动历史等密切联系,在隧道开挖前客观存在。在这种应力场中修建隧道就必须了解它的状态及其影响。岩体的初应力状态与施工引发的附加应力状态不同,它对坑道开挖后围岩的应力分布、变形和破坏有着极其重要的影响。可以说,不了解岩体初应力状态就不可能对隧道开挖后一系列力学过程和现象做出正确的评价。岩体的初应力状态通常受两类因素的影响:第一类型因素有重力、温度、岩体的物理与力学性质、岩体结构、地形及其他常规因素;第二类型因素包括突发性或局部性的。例如地壳运动、地下水活动、人类长期活动。

4.2.4 隧道开挖后的应力状态(Stress state after tunnel excavation)

The excavation of the tunnel removed part of the rock mass that was originally stressed in the tunnel, destroyed the equilibrium state of the initial stress field of the surrounding rock, and the surrounding rock changed from a relatively static state to a fluctuating state. The surrounding rock tries to reach a new balance, and its stress and strain start a new change movement. As a result of the movement, the stress of the surrounding rock redistributes and deforms to the excavated tunnel space, forming low-stress area, high-stress area, and original stress district three areas.

隧道的开挖,移走了隧道内原来受应力的部分岩体,破坏了围岩初始应力场的平衡状态,围岩从相对静态变为起伏状态。围岩力图达到新的平衡,在隧道开挖前就客观存在。运动的结果使围岩应力重新分布并向隧道内形变,形成低应力区、高应力区、原始应力区 3 个区域。

4.2.5 围岩压力的确定方法(Determination method of surrounding rock pressure)

(1) Direct measurement method: It is a practical method, and for tunnel engineering, it is also the direction of research and development; but due to the restriction of measurement equipments and technical level, it is not commonly used at present.

(2) Empirical method or engineering analogy method: Based on the statistics and summary of a large number of actual data of previous projects, the empirical value of surrounding rock pressure is proposed according to different surrounding rock classifications as the basis for determining surrounding rock pressure in subsequent tunnel projects. It is the most commonly used method at present.

(3) Theoretical estimation method: It is a method to study the surrounding rock pressure theoretically on the basis of practice.

(1) 直接量测法:是一种实用方法,对于隧道工程,也是研发的方向;但由于受量测设备和技术水平的制约,目前还不能普遍采用。

(2) 经验法或工程类比法:基于对以往项目的大量实际数据的统计和总结,按围岩的不同分类提出围岩压力的经验数值,作为后续隧道项目确定围岩压力的依据。这是目前最常用的方法。

(3) 理论估算法:是在实践的基础上从理论上研究围岩压力的方法。

4.3 隧道施工(Tunnel Construction)

4.3.1 钻爆隧道施工(Drilling-blasting method)

Drilling and blasting for tunnel construction can be used in geology ranging from hard rock with low strength, e.g. marl, loam, clay, gypsum, chalk, to the hardest rocks, such as granite, gneiss, basalt or quartz. Due to this large range of possible usage, drill and blast can be advantageous for very changeable ground conditions. The drill and blast work and the extent of the tunnel support can be adjusted with every heading advance if required.

隧道施工的钻探和爆破适用的地质范围,能从低强度硬岩,例如泥灰岩、壤土、黏土、石膏、白垩土,到最坚硬岩,例如花岗岩、片麻岩、玄武岩或石英。由于使用范围很广,钻爆法对多变的地质状况较为有利。如有必要,还可随每次掘进来调整钻孔、爆破作业和隧道支护范围。

1. 钻探(Drilling)

The first process involved in a blasting operation is the drilling. The principle of drilling is to obtain maximum penetration to enable placement of the explosive charges (Figure 4.3).The drilling system consists of the drill; the drill steel(or rod); and the bit. The force from the drill travels through the drill steel to the rock bit, penetrating the rock. Bits are designed for percussion, rotary, or both. The coordination of these forces with the slope, or the geometric properties of the bit, is what enables the bit to penetrate the rock. The rotation of the bit against the bottom of the borehole creates shear stresses in the rock, causing its separation. The percussion, or hammering, chips the rock by the combination of compressive and shear stresses created by the bit.

钻探是爆破操作的第一道工序。钻探的原则是获得最大贯入、能够来放置炸药。钻探系统由钻机、钎钢(或钻杆)和钻头组成,钻机的力通过钻杆传到钻头,以此贯入岩石(图 4.3)。钻头是为冲进、旋进或冲旋兼具来设计的。这些力与其斜坡协调或与钻头几何性质配合,这就是钻头能够贯入岩石的原因,钻头抵着钻孔底部旋转产生剪应力,导致岩石破碎。钻头冲击或锤击所产生的压应力和剪应力共同将岩石击成碎片。

图 4.3 隧道钻孔施工

(Figure 4.3 Tunnel drilling construction)

2. 爆破(Blast)

When the blast first occurs, the explosion creates a sudden application and quick release of high pressure, sending a shock wave through the rock from the borehole, after first crushing a small amount of rock immediately around the borehole. This compressive shock wave travels from the borehole through the entire rock mass as an elastic wave, with its velocity a function of rock density; the denser the rock is, the faster the wave travels. The wave travels the distance from the borehole to the nearest free face (this distance is called burden) and is reflected back to the borehole. If the shock wave reaches a change in density, a portion of the wave will return to the borehole and the remainder will continue through the different material in a weakened state (Figure 4.4).

当爆破首次发生时,爆炸产生的高压瞬时施加又迅即释放,在第一次粉碎紧围钻孔的少量岩石后,发出冲击波从钻孔穿过岩石。这种压缩冲击波以弹性波形式从钻孔穿透整个岩体,其速度是岩石密度的函数,岩石越致密,此波传播越快。波从钻孔传播到最近的自由面(此距离称为负载),又反射回钻孔。冲击波到达密度变化处时,一部分波将返回钻孔,其余部分将在减弱状态下继续穿越不同材质(图 4.4)。

图 4.4 隧道爆破施工

(Figure 4.4 Tunnel blasting construction)

4.3.2 隧道掘进机(Tunnel boring machine)

The tunnel boring machine (TBM) refers to a machine for excavating tunnels (in hard rock) with a circular full-cut cutter head, generally equipped with disc cutters.The rock is cut using these excavation tools by the rotation of the cutter head and the blade pressure on the face.

隧道掘进机(TBM)是指通常配备圆盘刀具,装有环形全切刀头在坚硬岩石中掘进隧道的机具。岩石就是利用这些挖掘工具,借助刀头旋转和刀面压力来切割破碎的。

TBMs exist in many different diameters, ranging from microtunnel boring machines with diameters smaller than 1m to machines for large tunnels, whose diameters are greater than 15m (the largest TBMs'diameters are now in excess of 19m,Figure 4.5). TBMs are available for many different geological conditions. Which means, for example, the type of tunnel face support required and the excavation procedure as well as numerous other technical requirements can be solved in many different ways. Every tunnel is different and hence there are often frequent technical advancements in this field. Although TBMs are often designed for specific projects, i.e. with a specific diameter in order to cope with certain ground conditions, these days refurbished machines are becoming more common and projects are actually designed around the machines available. An example of this is when the diameter of the new project is chosen to suit the old machine, only the cutter head would be redesigned for the specific ground conditions expected.

TBM 直径多种多样，从直径小于 1m 的微型隧道掘进机，到直径大于 15m 的大型隧道掘进机（目前最大的 TBM 直径超过 19m，图 4.5）。TBM 可用于多种不同地质条件。这意味着，譬如，所需的隧道工作面支护类型、开挖程序以及繁多其他技术要求，都能通过许多不同的方式来解决。每条隧道都存在差异性，因此在该领域屡屡涌现技术进步。尽管 TBM 往往是为特定项目而设计的，即为了应对某些地面条件而制成特定直径，但是如今翻新机越发普遍，且项目实际上围绕可用机来设计。一个例子是，当新项目所选直径选为适合旧机器时，仅针对预期的特定地面条件重新设计刀头。

图 4.5 隧道掘进机
(Figure 4.5 TBM)

Open face TBMs are mainly used in ground with a significant strength and stand-up time, but there are certain types of TBM which can be used in soft ground. Generally the main differences are the cutter tools, also known as cutter dressing, requirements on the cutter head and the face support are usually higher since the ground stability is generally lower.

敞开式掘进机主要用于高强度和自稳时间长的地层，但也有某些类型的掘进机能用于软土地面。一般来说，主要区别在刀具，又称修刀，由于地面稳定性一般较低，因此对刀盘和刀面支撑的要求较高。

For soft ground conditions different cutter tools from those discussed in the previous section can be used. Some examples of these different cutter tools are described in the following section. Hard rock cutter tools such as disc cutters can also be used in sandstone or limestone. Normally cutting tools work in soft ground by scraping material from the face. However, larger hard inclusions such as boulders, may occur under some circumstances. These inclusions have to be reduced in size before they can be quarried from the face. This is either done by hand, if the conditions at the face allow access and the inclusions are scattered, or by disc cutters. In addition, a stone crusher can be placed behind the cutterhead on slurry machines.

对于软土地层,能使用不同于前一节所讨论的刀具。下一节将介绍这些不同刀具的一些实例。硬岩石刀具,例如圆盘刀具,也能用于砂岩或石灰岩。正常情况,刀具在松软地面通过从表面刮削来工作的。然而在某些情况,可出现较大的硬包体,例如巨石。这些包体必须先削减尺寸,方能从工作面开采下来。当工作面条件允许进入并且块体分布零散时,可通过人工开采或圆盘刀具开采完成。此外,浆机刀盘后面还能放置一台碎石机。

4.3.3 沉管法(Immersed tube method)

Although a majority of tunnel construction occurs within the ground, for example TBM and NATM(new Austrain tunnelling method) tunnelling, there are techniques are constructed differently, notably immersed tube tunnels. An immersed or submerged tube tunnel is a type of cut-and-cover tunnel, but located underwater using pre-fabricated elements constructed in the dry at some distance from the tunnel location and made watertight with temporary bulkheads. These elements are then floated into position, lowered into a dredged trench on the river/sea bed, and joined together. One of the key design criteria for this type of tunnel, in contrast to bored tunnels, is the need to ensure adequate stability against uplift.

尽管大多数隧道施工发生在地下,例如 TBM 和 NATM(新奥隧道施工法)隧道,但也有一些技术的施工方式不同,最著名的要属沉管隧道。浸入或淹没管式隧道属于明挖隧道类型,但其位于水下,所用预制构件在距隧道一定距离的干燥处制成,并以临时隔板水密处理。然后将这些构件漂浮到河/海床上低于疏浚沟渠中适当的位置再连接在一起。与钻孔隧道相比,此类隧道的关键设计准则之一是要确保足够的抗隆起稳定性。

Immersed tube tunnels are ideal for crossing rivers and estuaries in urban areas. Due to the location just under the river/sea bed, this method can be considerably cheaper than excavating or boring (TBM driving) through the ground under the river/sea bed (Figure 4.6). The surface infrastructure (roads, railtrack) needs to be connected to the tunnel but is constrained by limits on gradients that are suitable for cars or trains. Therefore, the deeper the tunnel is, the longer this lead in section needs to be. As bored tunnels are generally constructed at a greater depth, which is necessary for ground stability during construction, they are often longer than immersed tube tunnels and hence more costly. Bored tunnels may also be technically more challenging due to the high water pressures, which can be a problem during construction.

沉管隧道是穿越城区河流和河口的理想选择。由于部位刚好在河床/海床之下,这比在河床/海床下方地层中开挖或钻孔(TBM 掘进)便宜了相当多(图 4.6)。地面基础设施(道路、轨道)需要连接到隧道,但受汽车或火车适宜坡度上限的制约。因此,隧道越深,需要的引入段越长。出于施工期间地层稳定性的需要,一般钻孔隧道修建的埋深更大,所以比沉管隧道更长,因而成本更高。由于钻孔隧道高水压,在技术上可能更具挑战性,这可能成为施工期面临的一个问题。

图 4.6 港珠澳大桥沉管隧道

(Figure 4.6 Immersed tube tunnel of Hongkong-Zhuhai-Macao Bridge)

The first immersed tube tunnel was built in the United States for conveying water across the Shirley Gut in Boston Harbor in 1894. The first transportation immersed tube tunnel was the Michigan Central Railroad Tunnel under the Detroit River in the United States which was completed in 1910.

1894 年,美国建造了第一条沉管输水隧道,横跨波士顿港狭窄的雪莉峡。第一条运输沉管隧道是于 1910 年建成的、位于美国底特律河下的密歇根中央铁路隧道。

4.4 专业词汇(Specialized Vocabulary)

cave body 洞身
cave gate 洞门
lighting device 照明器材
cave lining 洞身衬砌
straight-wall timbering 直墙式材砌
curved-wall lining 曲墙式衬砌
bias lining 偏压衬砌
rectangular-section lining 矩形断面衬砌
shotcrete lining 喷混凝土衬砌
anchor rod 锚杆
supporting structure 支护结构
integral lining 一体式衬砌

composite lining　复合式衬砌
shaft　n. 竖井
inclined shaft　斜井
parallel pilot pit　平行导坑
surrounding rock pressure　围岩压力
tunnel structure　隧道结构
drill and blast　钻爆
tunnel construction　隧道施工
hard rock　硬质岩
granite　n. 花岗岩
gneiss　n. 片麻岩
basalt　n. 玄武岩
quartz　n. 石英
the tunnel support　隧道支护
marl　n. 泥灰岩
loam　n. 亚黏土
clay　n. 黏土
gypsum　n. 石膏
chalk　n. 白垩岩
drill　vi. 钻孔
blast　n. 爆破
bit　n. 钻头
drill rod　钻杆
bored tunnel　钻孔隧道
penetrate　vi. 渗透，穿透，贯通
shear stress　剪应力
borehole　n. 钻孔
percussion　n. 撞击
compressive stresses　压缩应力
shock wave　冲击波
elastic wave　弹性波
rock density　岩石密度
tunnel boring machines　隧道掘机
disc cutter　盘刀
circular full-cut cutter head　圆形全切刀头
open face TBMs　敞开式掘进机
soft ground　软地基，软土地层
stability　n. 稳定性
sandstone　n. 砂岩

limestone n. 石灰岩
stone crusher 碎石机
slurry machine 浆水机
immersed tube tunnel 沉管隧道

习题(Exercises)

1. Translate the following sentences into Chinese.

(1) Due to this large range of possible usage, drill and blast can be advantageous for very changeable ground conditions.

(2) The force from the drill travels through the drill steel to the rock bit, penetrating the rock.

(3) The percussion, or hammering, chips the rock by the combination of compressive and shear stresses created by the bit.

(4) The wave travels the distance from the borehole to the nearest free face (this distance is called burden) and is reflected back to the borehole.

(5) Open face TBMs are mainly used in ground with a significant strength and stand-up time, but there are certain types of TBM which can be used in soft ground.

2. Translate the following sentences into English.

(1) 尽管大多数隧道施工发生在地下,例如 TBM 和 NATM 隧道,但也有一些技术,特别是沉管隧道,其施工方式是不同的。

(2) 使用这些挖掘工具,通过旋转刀头和刀面上的压力切割岩石。

(3) 这种压缩冲击波以弹性波的形式从钻孔穿过整个岩体,其速度是岩石密度的函数,岩石越致密,波传播越快。

(4) 一般来说,主要区别在于刀具,又称修刀,由于地面的稳定性一般较低,因此对刀盘和工作面的支撑要求较高。

凿井与掘巷工程
(Sinking and Driving Engineering)

5.1 立井与施工巷道(Vertical Shaft and Construction Roadway)

Primary and secondary horizontal openings play a major role in the development of a mine. This is well documented in statistics showing that out of the total length of development openings driven in 1980, 82% were horizontal openings, 16% were raises, and the rest were shafts, slopes, etc. As seen in Figure 5.1, Shaft is vertical passageway having direct access to the surface.

一级和二级水平巷道在矿区开发中起重大作用。据完备编档的统计数据表明,1980 年开掘的所有开发矿井中,82%为水平井,16%为天井,其余为立井、斜井等。立井是直接通往地表的垂直通道,如图 5.1 所示。

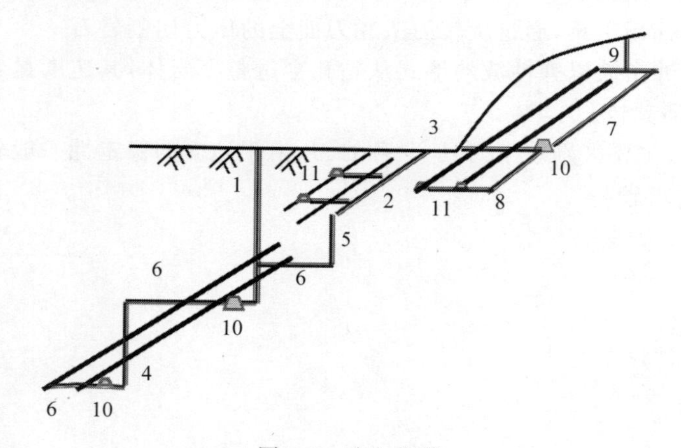

图 5.1 矿山巷道

1—立井;2—斜井;3—平硐;4—暗立井;5—溜井;6—石门;7—上山;8—下山;9—通风井;10—岩石平巷;11—煤层平巷。

(Figure 5.1 Mine openings)

1—shaft; 2—slope; 3—adit; 4—blind shaft; 5—draw shaft; 6—cross-cut; 7—rise; 8—dip; 9—ventilation shaft; 10—rock entry; 11—coal entry.

5.1.1 立井开挖法(Shaft mining)

Shaft mining or shaft sinking is excavating a vertical or near-vertical tunnel from the top down, where there is initially no access to the bottom.

立井开挖或凿井是从上往下挖掘一条垂直或接近垂直的隧道，最初这里完全没有通向底部的通道。

Shallow shafts, typically sunk for civil engineering projects differ greatly in execution method from deep shafts, typically sunk for mining projects. When the top of the excavation is the ground surface, it is referred to as a shaft; when the top of the excavation is underground, it is called a winze or a sub-shaft. Small shafts may be excavated upwards from within an existing mine as long as there is access at the bottom, in which case they are called "raises". A shaft may be either vertical or inclined (between 45° and 90° to the horizontal), although most modern mine shafts are vertical.

为土木项目开凿的浅立井与采矿项目开挖的深立井，施工方法有很大不同。当开挖顶面为地表时，称为立井；当挖掘的顶部在地下时，称为风井或副井。现有矿井内，只要底部有通道，就可向上挖掘出小立井，在这种情况下，这些小立井称为"天井"。立井可以是垂直的，也可以是倾斜的(与水平面呈 45°～90°)，不过大多数现代矿井都是垂直的。

If access exists at the bottom of the proposed shaft and ground conditions allow then raise boring may be used to excavate the shaft from the bottom up, such shafts are called borehole shafts. Shaft sinking is one of the most difficult of all development methods: restricted space, gravity, groundwater and specialized procedures make the task quite formidable.

当拟建的立井底部有通道且地面条件允许时，可采用天井钻探法从底部向上挖掘立井，这种立井称为钻孔立井。立井开凿是所有开发方法中最困难的：受限的空间、重力、地下水和专门的程序，使这项任务相当艰巨。

Historically those who had been among the most dangerous of all mining occupations and the preserve of mining contractors called sinkers. Today shaft sinking contractors are concentrated in Canada, Germany and South Africa.

历史上，凿井工是所有采矿职业中最危险的职业之一，并且是被称为"挖矿工"的采矿承包商的保护对象。如今，凿井承包商主要集中在加拿大、德国和南非。

5.1.2 矿井组成部分(Parts of a mine shaft)

As seen in Figure 5.2 and Figure 5.3, the most visible feature of a mine shaft is the head frame (or winding tower, poppet head or pit head) which stands above the shaft. Depending on the type of hoist used the top of the head frame will either house a hoist motor or a sheave wheel (with the hoist motor mounted on the ground). The head frame will also contain bins for storing ore being transferred to the processing facility.

如图 5.2、图 5.3 所示，矿井最明显的特点是井架(或卷绕塔、滑轮轴头或井口)矗立在矿井上方。根据所用提程式机的类型，井架顶部会安装提升机电机或滑轮（所配提升机电机安装在地面）。井架还包含用于储存待传送到加工设施的矿石的料仓。

图 5.2 德国马尔的废弃矿井

（Figure 5.2 Abandoned mine shafts in Marl，Germany）

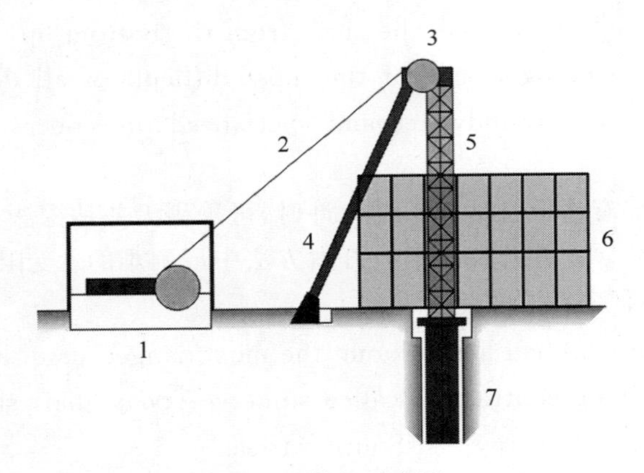

图 5.3 井架示意图

1—提升机；2—缆绳；3—滑轮；4—转向；5—伪边；6—提升机房；7—矿井。

（Figure 5.3 Schematic of head frame）

1—hoist；2—cable；3—wheel；4—sheer；5—false edge；6—hoist room；7—mine shaft.

At ground level beneath and around the headframe is the shaft collar，which provides the foundation necessary to support the weight of the headframe and provides a means for men，materials and services to enter and exit the shaft. Collars are usually massive reinforced concrete structures with more than one level. If the shaft is used for mine ventilation，a plenum space or casing is incorporated into the collar to ensure the proper flow of air into and out of the mine.

井架下方和井架周围的地面是井颈，它提供必要的基础来支撑井架的重量，并为人员、材料以及服务人员进出井筒提供一种手段。井颈通常是多层的厚实钢筋混凝土结构。当立井用于全矿通风时，在井口内加装一个充气室或套管，确保矿内空气正常流动。

At locations where the shaft barrel meets horizontal workings there is a shaft station which allows men, materials and services to enter and exit the shaft. From the drifts (drifts, galleries or levels) extend towards the ore body, sometimes for many kilometers. The lowest shaft station is most often the point where rock leaves the mine levels and is transferred to the shaft, if so a loading pocket is excavated on one side of the shaft at this location to allow transfer facilities to be built (Figure 5.4).

井筒与水平工作面的部位有一处井筒站，允许工作人员、材料和服务人员进出井筒。从平巷(巷道、坑道或水平面)延伸到矿体，有时长达数千米。最低的井站大多是在岩石离开矿层并传送到井筒的地方，这样的话，则在井筒的一侧挖出一个装卸箱，以便建造传送设施(图 5.4)。

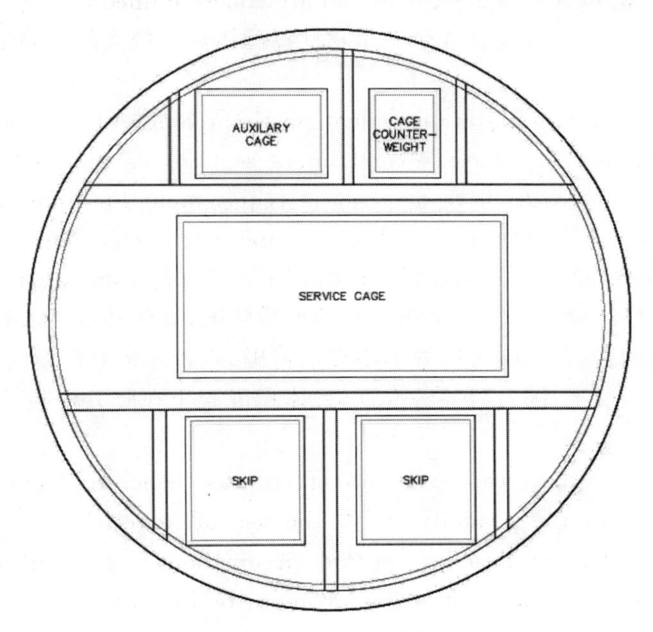

图 5.4　矿井平面示意图

(Figure 5.4　A plan-view schematic of a mine)

Beneath the lowest Shaft Station the shaft continues on for some distance, this area is referred to as the Shaft Bottom. A tunnel called a Ramp typically connects the bottom of the shaft with the rest of the mine, this Ramp often contains the mine's water handling facility, called the Sump, as water will naturally flow to the lowest point in the mine.

在最低的井站下面，井筒继续延伸一段距离，这个区域称为井筒底。一条叫作坡道的隧道通常将井底与矿井的其他部分连接起来，这个坡道常常包含矿水处理设施，称为水仓，因为水会自然地流到矿区最低点。

Shafts may be sunk by conventional drill and blast or mechaniszed means. The industry is gradually attempting to shift further towards shaft boring but a reliable method to do so has yet to be developed.

立井可采用常规钻孔、爆破或机械手段来开凿。该行业正逐步向钻井方向转移，不过尚未开发出可靠的方法。

5.1.3 井筒衬砌(Shaft lining)

The shaft lining performs several functions: it is first and foremost a safety feature preventing loose or unstable rock from falling into the shaft, then a place for shaft sets to bolt into and lastly a smooth surface to minimise resistance to airflow for ventilation.

井壁具有多种功能：首先是一种安全装置，防止松散或不稳定岩石落入立井；其次是立井安装锚固的地方；最后作为光滑表面将气流阻力降至最低，以实现通风。

In North and South America, smaller shafts are designed to be rectangular with timber supports. Larger shafts are round and are concrete lined.

在北美和南美，小的立井设计成带有木材支撑的矩形。较大的立井为圆形，采用混凝土衬砌。

Final choice of shaft lining is dependent on the geology of the rock which the shaft passes through, some shafts have several liners sections as required. Where shafts are sunk in very competent rock there may be no requirement for lining at all, or just the installation of welded mesh and rock bolts. The material of choice for shaft lining is mass concrete which is poured behind shaft forms in lifts of 6m as the shaft advances.

井筒衬砌的最终选择取决于所穿透岩石的地质状况，有些井按要求有多个衬垫，凡是井身开凿在完全合格的岩石之处，可以根本不要求衬砌，或者只装上焊接网和岩石锚杆。立井衬砌料选用大体积混凝土，随着竖井的推进，大体积混凝土在竖井模板后面浇筑，浇筑高度为6m。

Shotcrete, fiber, brick, cast iron tubing, precast concrete segments have all been used at one time or another. Additionally, the use of materials like bitumen and even squash balls have been required by specific circumstances. In extreme circumstances, particularly when sinking through halite, composite liners consisting of two or more materials may be required.

喷射混凝土、纤维、砖、铸铁管材、预制混凝土节段都曾一度用过。此外，在特定环境还要求使用沥青甚至挤压球。在极端环境，特别是在通过岩盐凿井时，可要求由多种材料组成的复合衬层。

The shaft liner does not reach right to the bottom of the shaft during sinking, but lags behind by a fixed distance. This distance is determined by the methodology of excavation and the design thickness of the permanent liner. To ensure the safety of persons working on the shaft bottom temporary ground support is installed, usually consisting of welded mesh and rock bolts. The installation of the temporary ground support is among the most physically challenging parts of the shaft sinking cycle as bolts must be installed using pneumatic powered rock drills.

在下沉过程中，轴衬不能直接到达轴的底部，而是滞后一个固定的距离。这个距离由开挖方法和永久衬层的设计厚度确定。为确保井底作业人员的安全,即安装临时地面支架,一般用焊接网和地脚螺栓。安装临时性地面支架是凿井周期中最具实际挑战性的部分,因为螺栓必须采用气动凿岩机来安装。

For this reason, and to minimize the number of persons on the shaft bottom a number of projects have successfully switched to shotcrete for this temporary lining. Research and development in this area is focusing on the robotic application of shotcrete and the commercialization of thin sprayed polymer liners.

因此,为了尽量减少井底的人员数量,许多项目成功地将临时衬砌改为喷射混凝土。该领域的研究和开发重点是喷射混凝土的机器人应用和薄喷射聚合物的商业化。

5.1.4 立井隔室(Shaft compartments)

Where the shaft is to be used for hoisting it is frequently split into multiple compartments by shaft sets, these may be made of either timber or steel. Vertical members in a shaft set are called guides, horizontal members are called buntons. For steel shaft guides, the main two options are hollow structural sections and top hat sections. Top hat sections offer a number of advantages over hollow structural sections including simpler installation, improved corrosion resistance and increased stiffness. Mine conveyances run on the guides in a similar way to how a steel roller coaster runs on its rails, both having wheels which keep them securely in place. Some shafts do not use guide beams but instead utilize steel wire rope.

凡是立井拟用于提升之处,往往以井架分成多个隔室,隔室由木材或钢材制成。井架中的垂直构件称为导柱,水平构件称为横撑。对于钢性导柱,两个主要可选项是空心结构段和顶帽段。顶帽段与空心段相比有许多优点,包括安装简便、耐腐蚀性改善、刚度增强。矿山输送工具在导轨上的运行方式类似于钢质过山车在导轨上运行。有些立井不使用导梁,而代之以导绳。

As seen in Figure 5.5, the largest compartment is typically used for the mine cage, a conveyance used for moving workers and supplies below the surface, which is suspended from the hoist on steel wire rope. It functions in a similar manner to an elevator. Cages may be single-deck, double-deck, or rarely triple-deck, always having multiple redundant safety systems in case of unexpected failure.

如图5.5所示,最大的隔室通常用于矿笼,一种用于将工人和补给运送到地表以下的运输工具,吊在钢丝绳上的绞车上。其功能类似于电梯。罐笼可以是单层、双层或三层的,带有多重冗余安全系统,以防意外和故障的发生。

The second compartment is used for one or more skips, used to hoist ore to the surface. Smaller mining operations use a skip mounted underneath the cage, rather than a separate device, while some large mines have separate shafts for the cage and skips. The third compartment is used for an emergency exit, it may house an auxiliary cage or a system of ladders. An additional compartment houses mine services such as high-voltage cables and pipes for transfer of water, compressed air or diesel fuel.

图 5.5 位于德国哈兹贝格堡典型矿笼
(Figure 5.5 Typical mine cage, located in Harzbergbau, Germany)

第二隔室用于一个或多个箕斗,用来将矿石提升到地面。小型采矿作业使用安装在罐笼下方的箕斗,而不是单独的器具,而一些大型矿山有单独的立井配罐笼和箕斗。第三隔室用作紧急出口,可装下辅助笼或多梯系统。还有一个补加的隔室来收纳矿区公用设施,例如高压电缆和管道,用于输送水、压缩空气或柴油燃料。

A second reason to divide the shaft is for ventilation. One or more of the compartments discussed above may be used for air intake, while others may be used for exhaust. Where this is the case a steel wall called a brattice is installed between the two compartments to separate the air flow. At many mines there are one or more complete additional separate "auxiliary" shafts with separate head gear and cages. It is safer to have an alternate route to exit the mine as any problem in one shaft may affect all the compartments.

划分立井的第二个理由是通风。上面讨论的隔室,一个或多个可用于进气,其余可用于排气。在这种情况下,在两个隔室之间安装了一堵称为隔板的钢墙,以分隔气流。许多矿中,补加一个或多个完备独立的"副"井,带独立的天轮和罐笼。更安全的做法是选用另一条备选路线退出矿区,因为一口井中的任何问题都会影响到所有隔室。

5.2 斜井与施工巷道(Inclined Shaft and Construction Roadway)

Slope mining is a method of accessing valuable geological material, such as coal or ore. A sloping access shaft travels downwards towards desired material. Slope mines differ from shaft and drift mines, which access resources by tunneling straight down or horizontally, respectively. In slope mining, the primary access to the mine is on an incline. Mine hoists may still be used to raise and lower loads on the incline if it is steep, but on shallower slopes, conveyor belts, locomotives or trucks may do the work. Drainage and ventilation of slope mines may be done using the primary slope, or it may be done using auxiliary shafts or bore-holes.

斜井采矿是一种取得有价值地质物资(例如煤炭或矿石)的方法。斜井身向下通向所需要的材料。斜井矿不同于立井矿和平巷矿,后两者分别通过直上直下或水平掘隧道来获取资源。斜井采矿中,通往矿井的主要通道是倾斜的。斜井陡峭时,仍可利用矿井提升机来提升或降低斜井的负荷,但对较浅的斜井,传送带、机车或卡车都可工作。斜井矿排水通风可利用主斜井,也可利用副井或钻孔。

5.2.1 优缺点(Merits and demerits)

The hanging wall of an inclined shaft is, of course, more difficult to support in weak ground than the walls of vertical shaft. Where vein material is soft and apt to cave, and the walls are strong, inclined shafts are usually put down in the foot wall side on the vein. Shafts so placed have the advantage over vertical hafts of being closer to loading chutes thus permitting shorter hauls.

当然,在软弱地层斜井悬挂壁比立井上壁更难支撑。如果矿脉材料柔软且易于塌陷,而且井壁坚固,则通常在矿脉的底部井壁侧设置斜井。这样的斜井的优点是比立井更接近装载槽,让拖运距离更短。

Experience has shown that, in general, the maintenance of inclined shafts materially exceeds that of vertical shafts. The reason for this may be that the inclined shafts are more often put down in soft rock and pass through faulted zones, whereas, the vertical shafts are usually in the firm wall rock.

经验表明,一般来说,斜井的维护费用大大超过立井。其原因可能是斜井往往建在软岩中并穿过断层带,而立井通常建在坚壁岩中。

However, in any well-constructed shaft, where the walls are reasonably firm, shaft maintenance is a relatively small item of the mining expense and the lower first cost of the inclined shaft might easily offset the extra cost of shaft repair.

然而,在任何建造良好的井中,井壁相当坚固,井筒维护费在采矿费用中是一个相对较小的款项,斜井的初始成本较低很容易抵消井筒维修的额外成本。

Inclined shafts of less than 60° dip are less common than those exceeding 60° dip, and often present special problems in sinking and equipping. They are apt to differ particularly in the timbering, laying of the rails, the type of skips employed and manner of loading and dumping. This discussion will, therefore, be confined to shafts having an inclination greater than 60° to the horizontal.

倾角小于60°的斜井比大于60°的要少,而且在凿井和配备中常常出现特殊问题。在木支架、轨道铺设、所用箕斗类型以及装卸方式诸方面相差悬殊。因此,此处讨论将限于水平倾角大于60°的斜井。

5.2.2 斜井尺寸和形状(Size and shape inclined of shaft)

Most inclined shafts are rectangular in crosssection and usually consist of two, three, four or more compartments arranged side by side on the line of the strike. Inclined shafts of three compartments are the most common, consisting of two hoisting compartments and one pipe and ladder compartment.

大多数斜井的横截面为矩形,通常由多间隔室组成,各隔室在走向线上并排开来。三室斜井最为常见,由两个提升室和一个管梯室组成。

The size of timbers required depends upon the load they are to bear. This is usually a rather indefinite factor in considering a new shaft, hence it is well to delay the installation of permanent timbers until some information concerning the tendency of the ground to stand or cave has been gained. When the size of timbers has been decided upon the full size of shaft outside of timbers is generally computed as the sum of compartments in the clear, plus the width of timbers, plus 4 inches outside of timbers in all sides.

木支架所要求尺寸取决于其必须承受的负荷。筹划新立井时,这通常是一个相当不明确的因素,因此,最好推迟安装永久性木支架,直到获得有关地面有直立或塌陷趋势的信息。决定了支架尺寸后,通常将支架外面立井的全尺寸计算为净空隔间尺寸,加上支架宽度,再加上支架外侧 4 英寸的总和。

5.2.3 分类(Classification)

1. 矿巷道按倾斜角分类(Mine openings classification according to the dip angle)

(1) vertical shaft and roadway;

(1) 立井与巷道;

(2) horizontal shaft and roadway;

(2) 水平井与巷道;

(3) inclined shaft and roadway.

(3) 斜井与巷道。

2. 斜井-与地面直接相通的斜井(Slope-inclined shaft which has direct access to the surface)

(1) Main slope;

(1) 主斜井;

(2) Auxiliary slope.

(2) 副斜井。

3. 斜井巷道(Inclined roadway)

(1) According to location: ①Rock rise or dip; ②Coal rise or dip.

(1) 按层位分: ①岩石上下山; ②煤层上下山。

(2) According to usage: ①Coal haulage rise or dip; ②Material transporting rise or dip; ③Return-air rise or dip; ④Men-walking rise or dip.

(2) 按用途分: ①运输上下山; ②运料上下山; ③回风上下山; ④行人上下山。

5.2.4 斜井施工工具(Inclined shaft construction tool)

Using tunnel boring machine (TBM) to construct inclined shafts in coal mines has become the main way for large-scale coal mines to enter deep mining faces.

采用隧道掘进机(TBM)构筑斜井已成为大型煤矿进入深部采煤面的主要方式。

During tunnelling several difficult geological conditions were encountered mainly including stress-induced spalling, discontinuity-controlled block failures, karst voids filled with sediments which were not predicted to that extent. In horizontal tunnelling these conditions are better manageable whereas in inclined shaft they are major challenges.

巷道掘进期间,曾遇到几种棘手的地质状况,主要是:应力诱发的剥落,间断受控块体破损,以及由沉积物而无法预测其所达范围的岩溶孔隙。在水平掘进中,这些状况较好掌控,在斜井中却都是重大挑战。

The inclined shaft excavation using a Tunnel Boring Machines (TBM) proved as a safer, fast and more economical, despite of the logistical difficulties and geological conditions, as compared to other possible methods (drill and blast, raise drill). This excavation method also allowed negotiating even larger geological problems.

尽管后勤保障困难和地质状况不良,但比起其他可能方法(钻爆法、升孔法)相比,采用掘进机(TBM)开挖斜井证明是安全、快速、更经济的。这种挖掘办法还可以解决更大的地质问题。

5.3 巷道的安全和通风(Safety and Ventilation of Roadway)

The design of underground mining operation requires the integration of transportation, ventilation, ground control, and mining methods to form a system which provides the highest possible degree of safety for mine personnel.

设计地下采矿作业的要求将运输、通风、地面控制和采矿方法综合起来,形成一个系统,为矿区人员提供尽可能高的安全程度。

5.3.1 安全和通风(Safety and ventilation)

Although ventilation, ground control, drainage, power, communications and lighting are auxiliary operations in the mine, definite decisions must be made about transportation, maintenance, ventilation, drainage, delivery of the supplies and power distribution in the design in order to ensure the safety of mine personnel.

通风、地面控制、排水、动力、通信以及照明,虽然都是矿区的辅助作业,但在设计中必须对运输、维护、通风、排水、补给的交付、配电做出明确的决定,以确保矿区人员的安全。

Mine production system mainly includes coal haulage system, ventilation system, material and refuse transportation system, drainage system, power supply system, communication and monitoring system, etc (Figure 5.6). Only the efficient and orderly operation of all systems can ensure the safety of the mine. The following is a brief introduction of each system.

矿区生产系统主要包括：煤炭拖运，通风，材料与废料运输，排水，供电，通信监控等诸多系统(图 5.6)。只有各系统都高效、有序运行，方能确保矿区安全。下面是对每一系统的简要介绍。

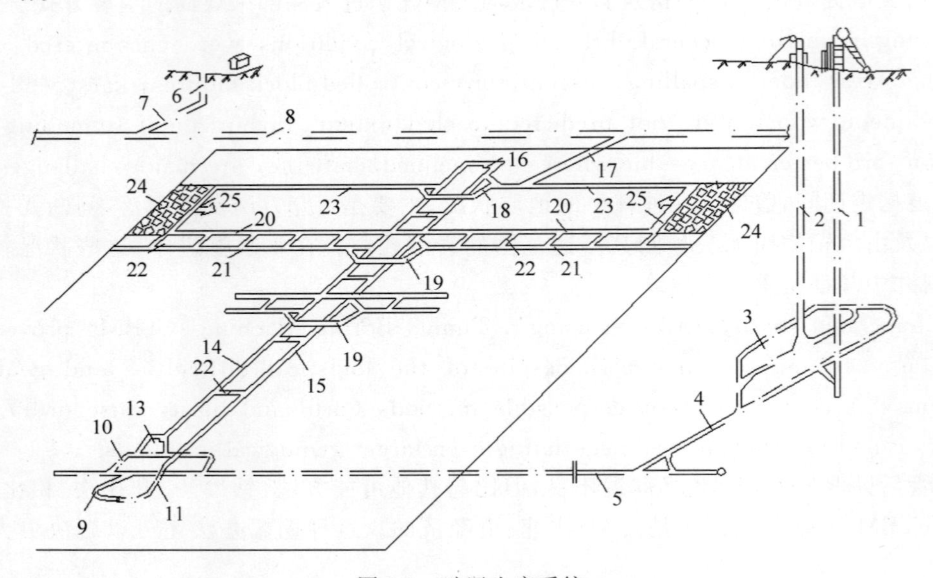

图 5.6 矿区生产系统

(Figure 5.6 Mine production system)

1. 运煤系统(Coal transportation system)

Coalface25→head entry20→haulage rise14→coal bunker12→district bottom station10→main haulage roadway5→main haulage crosscut4→pit bottom3→main shaft1→surface.

采煤面 25→区段运输巷 20→拖运上山 14→采区煤仓 12→采区下部车场 10→水平拖运大巷 5→主拖运石门 4→井底车场 3→主井 1→地面。

2. 通风系统(Ventilation system)

Fresh air: surface→auxiliary shaft2→pit bottom3→main haulage crosscut4→main haulage roadway5→district bottom material station11→track rise15→head entry21→coalface25.

新鲜空气：地面→副井 2→井底车场 3→主拖运石门 4→水平拖运大巷 5→采区下部材料场 11→采区轨道上山 15→下区段回风平巷 21→采煤面 25。

Dirty air: Coalface25→tail entry23→haulage rise14→district top station→district return-air crosscut17→main return-air roadway8→main return-air crosscut7→air shaft6→surface.

污浊空气：采煤面 25→区段回风平巷 23→拖运上山 14→采区上部山→采区回风石门 17→回风大巷 8→回风石门 7→风井 6→地面。

3. 排水系统(Drainage system)

Drainage system is opposite to the intaking air direction.

排水系统与进风气流方向相反。

Coalface25→head entry20→track rise→district bottom material station11→main haulage roadway5→main haulage crosscut4→pit bottom3→auxiliary shaft2→surface.

采煤工作面25→区段运输平巷20→采区轨道上山→采区下部车场11→水平运输大巷5→主运输石门4→井底车场3→副井2→地面。

4. 运料排矸系统(Material and refuse transportation system)

1) 运料系统(Material transportation system)

surface→auxiliary shaft2→pit bottom3→main haulage crosscut4→main haulage roadway5→district bottom station11→track rise15→tail entry20→Coalface25.

地面→副井2→坑底车场3→主运输石门4→水平运输大巷5→采区运输石门9→采区下部材料车场11→采区轨道上山15→区段回风平巷20→采煤工作面25。

2) 排矸与运料方向相反(Refuse transportation is opposite to the material transportation)

5. 动力供应系统(电源与压缩空气)(Power supply system(electric power and compressed air))

surface substation→central substation in pit bottom→district substation→district sublevel.

地面变电所→井下中央变电所→采区变电所→工作面配电(区段)。

Surface compressed-air house→auxiliary shaft→pit bottom→main haulage crosscut→main haulage roadway→working face.

地面压缩空气站→副井→井底车场→主要运输石门→主要运输大巷→工作面。

6. 通信监测系统(Communication and monitoring system)

Surface control center→auxiliary shaft→pit bottom→main haulage crosscut→main haulage roadway→working face.

地面调度中心→副井→井底车场→主要运输石门→主要运输大巷→工作面。

Surface monitoring house→auxiliary shaft→pit bottom→main haulage crosscut→main haulage roadway→mains tation→substation →monitoring area.

地面监测室→副井→井底车场→主要运输石门→主要运输大巷→主站→分站→监测区。

5.3.2 奎切克矿山救援(Quecreek mine rescue)

The Quecreek Mine rescue took place in Somerset County, Pennsylvania, when nine miners were trapped underground for over 77 hours, from July 24 to 28, 2002. All nine miners were rescued eventually (Figure 5.7).

2002 年 7 月 24—28 日，宾夕法尼亚州萨默塞特县发生了奎克雷克矿难救援事件，9 名矿工被困井下超过 77 小时，最终全部获救(图 5.7)。

图 5.7 带封顶救援孔现场(中间)，2013 年 7 月

(Figure 5.7 Site, with capped rescue hole (center), in July 2013)

The structural geology of the area caused the flooded mine void of the shallower Saxman Mine to be at a higher elevation than the active Quecreek Mine.

该地区的地质构造导致较浅的萨克斯矿被淹没的矿穴比活跃的奎切克矿更高。

At approximately 9 p.m. on Wednesday, July 24, the eighteen miners were in danger 240 feet underground, below the fields of Dormel Farm when the flooded Saxman mine was breached as the mining progressed eastward. Water had broken through the face and was inundating the entry, and the nine miners in the 1-Left panel area used the mine's phone system to notify the other group of nine miners in the 2-Left panel to evacuate immediately. These miners were able to escape at around 9: 45 p.m. and alert others, and a 911 call was made at 9:53 p.m. However, the mine was flooding too rapidly for the miners in the 1-Left panel area to evacuate. Twice they tried to travel in the four-foot-high tunnels over 3000 feet to a shaft that would lead them to the surface, but these were also flooded. Water continued to rise in the mine during the morning hours of Thursday, July 25.

7 月 24 日星期三晚上 9 点左右，这 18 名矿工在多莫尔农场地下 240 英尺处遇险，当时，随着采矿向东推进，水淹没的萨克斯曼煤矿被炸开。水已经冲破了工作面，淹没了入口，左 1 面板区域的 9 名矿工使用矿区电话系统通知左 2 面板区域的另一组 9 名矿工立即撤离。这些矿工在晚上 9 点 45 分左右得以逃生，并向其他人发出警报，9 点 53 分拨打了 911 电话。然而，矿区水浸得太快，左侧面板区域的矿工无法撤离。他们两次试图在长度超过 3000 英尺长、4 英尺高的隧道中行驶到一个竖井，这个竖井会把他们引到地表，但这些隧道也被淹没了。7 月 25 日星期四上午,矿区水位继续上升。

With the mine portal entrances to Quecreek mine nearly under water, rescue operations started immediately. While pumping water would begin at all mine locations and any nearby residential and commercial water wells, the mine rescue first focused on getting air to the trapped miners. With the help of Bob Long, an engineer technician for Civil Mining Environmental Engineering, GPS measurements were made and a 6.5-inch-diameter borehole was begun at 2: 05 a.m. The borehole was drilled to allow air to be pumped into the mineshaft where the miners were presumed to be, at the most up dip location near where the Saxman mine was breached. A four-member team started working about 3: 15 a.m. Thursday, and its drill cracked through what turned out to be 240 feet of rock, and into the mine shaft 1 hour and 45 minutes later. On July 25, at 5: 06 a.m., approximately 8 hours after the breakthrough, the 6.5in. hole was drilled into the mine. The drilling rig's air compressor pushed air into the mine, and the air returning from the borehole showed a marginal air quality of 19.3 percent oxygen(Figure 5.8).

由于通往奎切克矿的矿区桥门入口几乎被水淹没,救援行动立即展开。在所有矿区部位以及附近的住宅和商业水井开始抽水的同时,矿井救援首先关注为被困矿工提供空气。在土木采矿环境工程技术人员 Bob Long 的帮助下,进行了 GPS 测量,并于凌晨 2 点 5 分开始钻一个直径 6.5 英寸的钻孔。钻孔是为了将空气抽到据推测矿工所在的矿井区,位于萨克森矿破裂处附近最上倾部位。周四凌晨 3 点 15 分左右,一个 4 人小组开始工作,钻通了 240 英尺的岩石,并在 1 小时 45 分钟后进入矿井。7 月 25 日早上 5 点 06 分,在突破大约 8 小时后,钻出一个 6.5 英寸的洞。钻机的空气压缩机将空气推入矿井,从钻孔返回的氧气质量低于 19.3%(图 5.8)。

图 5.8 钻孔直径 6 英寸,深 240 英尺的通风井

(Figure 5.8 The 6 in. diameter, 240 ft deep air shaft drilled)

However, while the drilling rig's compressed air rapidly increased the oxygen content of the mine air, monitors showed the rising water was approaching 1825 feet above sea level, and rescuers feared they had perhaps an hour before the area where the miners had taken refuge would be under water. Mine ventilation expert John Urosek, of the U.S. Department of Labor's Mine Safety Health Administration, proposed creating a pressurized air pocket for the miners. The drill operator then used his rig's air compressor to pump and maintain 920 cubic feet per minute at a temperature of 197℉ (92℃).

然而,当钻机的压缩空气迅速增加了矿井空气含氧量时,监测器显示,水位正在上升,接近海平面以上1825英尺,救援人员担心,矿工们避难的地区会处于水下大概一小时。美国劳工部矿山安全健康管理局的矿井通风专家约翰·乌罗塞克提议为矿工们制一个压缩空气袋。然后,钻机操作员使用钻机的空气压缩机在197℉(92℃)的温度下泵送,并保持920立方英尺/分钟。

Meanwhile, an ongoing battle was to dewater the Quecreek Mine to allow rescue operations to be planned. Millions of gallons of water had to be pumped from the flooded coal mines as the water level needed to be lowered to prevent the loss of the air pocket in the mine area where the nine miners would congregate. Should a rescue hole penetrate the mine, the air pocket could escape and the air-filled void area become flooded, and the miners would drown. The second grave concern was the quality of the air in the mine.

与此同时,在奎切克矿正在进行一场排水战斗,让救援按计划行动。为了防止9名矿工聚集区气穴损失,必须降低水位,从被淹各煤矿中抽出数百万加仑的水。一旦救援孔贯入矿井,气穴就会逃逸,充满空气的孔隙区就会淹没,矿工会被淹死。第二个严重问题是矿内空气质量。

Immediately after the rescue hole penetrated the mine, all equipment was shut down in order to take an accurate relative air pressure reading between the mine and surface atmospheres. The pressure reading was zero, indicating that the pressures were equal and that the airlock would not be required. The compressor was turned off and the drill steels were removed from the 6in. hole. At 10:53 p.m. a special pen-shaped, two-way communication device was lowered into the 6in. air pipe, with a child's glow stick attached to it for visibility in the dark mine. Communication was established with the miners who confirmed that all nine were alive and well, except for the foreman who was experiencing chest pains.

救援孔贯入矿区后,立即关闭所有设备,以便在矿区与地表大气之间获取准确的相对气压读数。压力读数为0,表明压力相等,就不需要气闸。关闭压缩机,从6英寸的孔中取出钻钢。晚上10点53分,一个特殊的笔形双向通信装置被送入6英寸的空气管中,为了在黑暗矿内看清,管子上还附着一个儿童荧光棒。与矿工取得了联系,确认除领班曾经胸痛外,其余9名矿工均安然无恙。

At 12:30 a.m. on July 28, the 8 foot high steel mesh escape capsule, with supplies, descended into Rescue Hole No. 1, into the void where the men had languished in fear and hope for 77 hours. The miners were brought up in 15-minute intervals, and all nine miners were on the surface at 2:45 a.m(Figure 5.9).

7月28日中午12:30,8英尺高的钢网逃逸舱携带补给进入1号救援洞,进入那个人们

图 5.9 救援场地的“生命纪念碑”

(Figure 5.9 The “Monument for Life” at the rescue site)

在恐惧和希望中煎熬了 77 小时的洞中。此后,每隔 15 分钟就会有矿工升井,9 名矿工于凌晨 2 点 45 分全部被营救到地面(图 5.9)。

5.4 专业词汇(Specialized Vocabulary)

shaft n. 竖井;通风井

roadway n. 巷道

mine n. 矿井

horizontal opening 水平井

raise n. 上升井

shaft mining 立井采矿(法)

civil engineering 土木工程

shallow shaft 浅井

deep shaft 深井

mine shaft sinking 矿井开凿

head frame 井架

hoist n. 升降机

cable n. 缆绳

shotcrete n. 喷浆混凝土

fiber n. 纤维

brick n. 砖

cast iron tubing 铸铁管材

precast concrete segment 预制混凝土段

bitumen n. 沥青

slope n. 斜井
blind shaft 暗立井
draw shaft 溜井
ventilation shaft 风井
rock entry 岩石平巷
coal entry 煤层平巷
coalface n. 采煤面
production n. 生产;产量
coal haulage system 运煤系统
ventilation n. 通风
ventilation system 通风系统
fresh air 新鲜空气
dirty air 乏风,污风
working face 工作面
production system 生产系统
drain v. 排水
drainage system 排水系统
drive v. 掘进
hoist v. 提升
ground control 地面控制
communication n. 通信
lighting n. 照明

习题(Exercises)

1. Translate the following sentences into Chinese.

(1) Primary and secondary horizontal openings play a major role in the development of a mine.

(2) This is well-documented in statistics showing that out of the total length of development openings driven in 1980, 82% were horizontal openings, 16% were raises, and the rest were shafts, slopes, etc.

(3) Slope mines differ from shaft and drift mines, which access resources by tunneling straight down or horizontally, respectively.

(4) Drainage and ventilation of slope mines may be done using the primary slope, or it may be done using auxiliary shafts or bore-holes.

(5) The design of underground mining operation requires the integration of transportation, ventilation, ground control, and mining methods to form a system which provides the highest possible degree of safety for mine personnel.

2. Translate the following sentences into English.

(1) 土建工程常用的浅立井与采矿工程常用的深立井在施工方法上有很大的不同。

(2) 如今,立井承包商主要集中在加拿大、德国和南非。

(3) 斜坡开采是一种获取有价值的地质物资(例如煤炭或矿石)的方法。

(4) 所需木材的尺寸取决于它们所能承受的重量。

(5) 第二个严重的问题是矿井里的空气质量。

边坡工程
(Slope Engineering)

The slope is one of the most basic naturalgeological environments for human survival and the project activities. Rock engineering of slopes on facilities such as railways, highways and power facilities must have a high degree of reliability and safety for the public. This requires that they be stable against both large-scale failure and rock falls.

边坡是人类生存和工程活动最基本的自然地质环境之一。铁道、公路、电力等设施的边坡岩体工程,必须对公众保证高度可靠性和安全性。这就要求边坡能够稳定地抵抗大规模的破坏和崩塌。

Failures of rock slopes, both man-made and natural, include rock falls, overall slope instability and landslides, as well as slope failures in open pit mines. The consequence of such failures can range from direct costs of removing the failed rock and stabilizing the slope to possibly a wide variety of indirect costs. Examples of indirect costs include damage to vehicles and injury to passengers on highways and railways, traffic delays, business disruptions, loss of tax revenue due to decreased land values, and flooding and disruption to water supplies where rivers are blocked by slides. In the case of mines, slope failures can result in loss of production together with the cost of removal of the failed material, and possible loss of ore reserves if it is not possible to mine the pit to its full depth.

岩石边坡破损,无论是人为的还是自然的,都包括:岩石崩落,边坡总体失稳和滑坡,以及露天矿的边坡破损。这些破坏的后果包括清除破损岩石和稳固斜坡的直接成本和名目繁多的间接成本。间接成本的实例有:车辆受损,公路铁路乘客受伤,交通延误,业务中断、土地贬值造成的税收损失,以及河流因滑坡堵塞所致的洪水泛滥和供水中断。就矿区而言,边坡破损能导致生产损失连带除去受损材料的费用,以及矿坑无法开采到最大深度,可能会造成矿石储量的损失。

Geology, nature of slope materials and topography are always important factors in the case of natural slopes and often in the case of man-made slopes. Often some idealization of problems is required for clarity and for the effective application of basic concepts.

对于自然边坡和人工边坡,地质、边坡材料自然性质和地形始终是重要因素。为了清晰和有效地应用基本概念,常常需要对问题进行一些理想化。

6.1 自然边坡(Natural Slopes)

The slopes formed due to natural process and exist naturally are called natural slopes (Figure 6.1). Natural slopes are those that exist in nature and are formed by natural causes. Such slopes exist in hilly areas.

由于自然过程而形成并自然存在的斜坡称为自然边坡(图 6.1)。自然边坡是自然界中存在的、由自然原因形成的边坡。这种边坡出现在丘陵地区。

图 6.1 自然边坡

(Figure 6.1 Natural slopes)

Natural slopes in soil and rock are of interest to civil and mining engineers, engineering geologists, applied geomorphologists, soil scientists, and environmental managers. The material composing any slope has a natural tendency to slide under the influence of gravitational and other forces (those due to tectonic stresses, seismic activity etc.) which is resisted by the shearing resistance of the material. Instability occurs when the shearing resistance is not enough to counterbalance the forces tending to cause movement along any surface within a slope. Natural slopes which have been stable for many years may suddenly fail due to one or more of the following main causes:

土壤和岩石中的自然边坡一直吸引土木和采矿工程师、工程地质学家、应用地貌学家、土壤科学家和环境学家的兴趣。构成任何边坡的材料,在重力和其他力(构造应力、地震活动等)的影响下均有自然滑动的倾向,而这些力受到材料剪切阻力的抵抗。当抗剪阻力不足以抵消趋于造成沿斜坡内任何表面移动的力时,就出现失稳。多年来一直稳定的自然边坡可由于以下一种或多种主要原因而突然崩塌:

(1) External disturbance in the form of cutting or filling of parts of a slope or of ground adjacent to it resulting in an alteration of the balance between forces tending to cause instability and forces tending to resist it.

(1) 以切割或填充其几部分的斜坡或其附近地面的形式产生的外部干扰,导致倾向于引起不稳定的力与趋于与之抵抗不稳定力之间的平衡发生改变。

(2) External disturbance in the form of seismic activity (earth tremors or earthquakes).

(2) 地震性活动(地球震颤或地震)形式的外部干扰。

(3) Increase of pore water pressures within a slope (e.g. rise in water table) due to significant changes in the surrounding areas such as deforestation, filling of valleys, disturbance of natural drainage characteristics, urbanization, construction of reservoirs, and exceptional rainfall.

(3) 由于周围地区的显著变化,例如砍伐森林、填满山谷、扰乱自然排水特性、城市化、兴建水库、雨量异常等,导致斜坡内的孔隙水压力增大(例如水位上升)。

(4) Increase of pore water pressures to equilibrium values several years after a cutting (in slope material of low permeability) which resulted in significant post-construction decrease of the pore water pressures below their equilibrium values.

(4) (低渗透边坡材料)开挖数年后孔隙水压力增大至平衡值,导致施工后孔隙水压力显著降低至其平衡值以下。

(5) Progressive decrease in shear strength of slope materials: This may be due to significant deformations which do not appear to constitute instability but lead to it. Such deformations may occur due to sustained gravitational forces and slope disturbances of an intensity not high enough to cause complete failure. Deformations often occur along major natural discontinuities, ancient slip surfaces and tectonic shear zones within a slope.

(5) 边坡材料抗剪强度逐渐降低:这可由明显的变形造成,这些变形看似不至于导致稳定,但却导致不稳定性。这类变形可能是由于持续性重力和强度不足以引起完全损坏的斜坡扰动共同造成的。变形经常发生在斜坡内的主要的自然不连续面、古滑坡滑动面和构造剪切带上。

(6) Progressive change in the stress field within a slope: Every natural geological formation has an "initial" stress field which may be significantly different from one considered in terms of the weight of the material alone. Lateral stresses may occur which do not bear any predictable relationship with the vertical stress computed from gravitational considerations. The unique "initial" stress field of any slope depends on its geological background and other natural factors. The stress history of the slope materials is of tremendous importance. Attempts have been made in recent years to develop methods for the prediction of initial stresses in soils on the basis of laboratory tests. However, it is recognized that reliable information is best obtained from in-situ measurement in soil and rock. In some cases, these measurements present considerable difficulties.

(6) 边坡内应力场的渐进变化:每一自然地质构造都有一个"初始"应力场,这与单考虑材料重量时的应力场可存在显著差异。可出现侧向应力,而这些力与考虑重力算出的垂直应力没有任何可预见的关系。任何边坡独特的"初始"应力场取决于其地质背景和其他自然因素。边坡材料的应力史极其重要。近年来,人们试图在实验室试验的基础上研发出预测土壤初始应力的方法。然而,人们认识到可靠的信息最好从对土壤岩石的现场测量中获得。在某些情况下,这些测量存在相当大困难。

A change in the initial stress field may occur due to causes similar to those which produce a progressive decrease of shear strength. Release of stresses may accompany or follow most forms of slope disturbance. Often this leads to changes in both the magnitude and orientation of the stresses.

初始应力场的变化可由于类似于导致剪切强度逐渐降低的原因造成。应力释放可伴随或跟随大多数形式的边坡扰动。这常导致应力的幅值和取向的变化。

(7) Weathering: It is now widely recognized that weathering may occur at a rate rapid enough to be of concern in the design of engineering works. Therefore, it is important to consider not only the existence of weathering which has occurred in the past but also the possibility of continued and even accelerated weathering. Weathering of soils and rocks destroys bonds and reduces shear strength. The weathering of over consolidated clays (clays which have experienced a higher overburden pressure in their past than their present overburden pressure) and shales increases their recoverable strain energy and consequently their capacity for progressive failure. This occurs due to the destruction by weathering of diagenetic bonding in these materials. Weathering may be accelerated by slope disturbance and by exposure to atmospheric and other agencies such as stream action.

(7) 风化作用:现在普遍认为,风化发生率可以快得足以引起工程设计的关注。因此,重要的是不仅要考虑过去已经发生的风化,还要考虑持续式甚至加速式风化的可能性。土壤和岩石的风化破坏了黏结,并降低了抗剪强度。超固结黏土(其过去历经的超负载压力高于现在)和页岩的风化增加了两者的可恢复应变能,进而且渐进损坏能力。正是由于这些材料中的成岩胶合风化作用而造成的损坏。边坡扰动、暴露于大气以及其他能动因素(例如水流作用)均可加速风化。

It must be emphasized that many failures of natural slopes are imperfectly understood and that there may be other critical factors which influence the long-term stability of natural slopes. In rocks a slow and cumulative process of deterioration and destruction depending on climatic factors is always at work and thousands of years may elapse before a slope fails. The following factors are of primary importance in the time-dependent process leading to rock slides:

必须强调的是,对自然边坡的许多损坏还不完全了解,还有其他一些关键因素可影响自然边坡的长期稳定性。在岩石中,气候因素导致的缓慢累积和恶化破坏过程总在持续作用,在边坡崩塌之前,可能要持续数千年之久。在导致岩石滑坡的随时间变化的进程中,以下因素是最重要的:

(1) The presence of a system of joints known as valley joints along which rock masses are often detached during slides. The occurrence of large residual stresses in natural formations is well-known and it is believed that valley joints are often formed by further stress changes and uneven deformations which occur during the formation of valleys. The release of strain energy stored due to large overburden in previous geological periods has thus an important role to play in the development of such joint systems. Each system at a

particular depth is associated with a different stage of erosion in a rock mass.

(1) 存在一种称为山谷节理系的节理系统，滑坡期间，岩体常沿着此种节理系脱离。众所周知，天然地层中存在大残余应力，又认为山谷节理系往往是在山谷形成期间应力的进一步变化和参差变形形成的。因此，在以往地质期，覆盖层过大而所储存应变能释放出来，对此类节理系统的发育起着重要的作用。特定深度的每一个系统都与岩体的不同侵蚀阶段关联。

(2) The presence of residual ground stresses which have still not been relieved during the formation of valley joint systems.

(2) 山谷节理系统形成期间，残余地应力的出现仍未得到缓解。

(3) The presence of water in open joints of rock masses which influences stability by exerting a direct outward force as well as by decreasing the effective stress and hence the shear strength on failure surfaces.

(3) 岩体裂隙中出现的水，通过直接施加外向力，来降低破裂面上的有效应力和抗剪强度，进而影响稳定性。

(4) Fluctuation of water pressure in a joint system causes cumulative opening of joints during periods of high pressure following precipitation. The wedging of crushed rock in joints often prevents them from returning to their original position after opening under high water pressures. Fatigue failure may also result due to fluctuations of pressure leading to further extension of open joints through intact rock. Thus gradually, the proportion of a potential failure plane passing through jointed rock is increased to a critical value. At such a stage further decrease of the area of intact rock is no longer consistent with slope stability.

(4) 节理系统中的水压波动，在降水随后的高压期间导致节理系的累积性张开。节理系中的碎石楔入常防止其在高水压下张开后返回原来位置。疲劳破损也可能因压力起伏导致开放式节理系进一步延伸穿过未能动岩石所造成。因此，通过带节理岩石的潜在损坏面比例逐渐增加到临界值。在这一阶段，未能动岩石区的进一步减少不再与边坡稳定性一致。

The foregoing remarks were made in the context of hard rock slopes. However, there are similar processes at work in soil slopes. The cumulative influence of natural processes on long-term stability can rarely be quantified. In many instances concerning both soil and rock slopes a significant level of uncertainty exists with regard to stability and this has beenemphasized.

上述评述是在硬岩斜坡的环境中提出的。然而，在土坡上也有类似的运作过程。自然过程对长期稳定性的累积影响很难量化。在许多关乎土壤和岩石两种边坡的事例中，在稳定性方面存在很大程度的不确定性，这一点已经强调过。

It is most important to draw a clear distinction between natural slopes with or without existing slip surfaces or shear zones. A knowledge of the existence of old slip surfaces makes it easier to understand or predict the behavior of a slope. Such surfaces are often a result of previous landslide or tectonic activity. Shearing surfaces may also be caused by other processes including valley rebound, glacial shove, periglacial phenomena such as solifluction and nonuniform swelling of clays and clay-shales. It is not always easy to recognise landslide areas (while post-glacial slides are readily identified, preglacial surfaces may lie buried beneath glacial sediments) or locate existing shear surfaces on which previous movements have occurred. However, once pre-sheared strata have been located, evaluation of stability can be made with confidence.

最重要的是在带或不带滑动面或剪切带的两种天然斜坡之间划清界限。了解旧滑动面的存在可以使理解或预测斜坡性态变得更加容易。这些地表常常是先前滑坡或构造活动的结果。剪切面也可由其他过程引起的,包括山谷反弹,冰川铲动,冰缘现象例如黏土和黏土页岩的溶解和不均匀膨胀。识别滑坡区(虽然很容易辨识冰期后的滑坡,但冰期前的地表可以埋在冰川沉积物下)或定位以前发生过运动的现有剪切面并非总是很容易。然而,一旦定位出预剪切地层,就能很有把握进行稳定性评价。

There are two main reasons for this: ①renewed movements in these slopes are likely to occur along existing slip surfaces and ② residual strengths operative on such surfaces can be determined or inferred with greater certainty than shear strengths of unsheared, *in situ* soils and rocks. On the other hand, The evaluation of stability of slopes with no previous landslide activity and no existing shear surfaces is a far more difficult task.

这有两个主要原因:①这些边坡可能沿现有滑动面重新移动;②与未开挖的原位土壤和岩石的抗剪强度相比,能以更大把握确定或推断作用在此类地表上的残余强度。另一方面,对既往无滑坡活动又无剪切面的边坡,评价其稳定性是一项越加艰难的任务。

6.2 人工边坡(Man-made Slopes)

Man-made slopes are formed by humans as per requirements. The slopes formed by unnatural process. The sides of cuttings, the slopes of embankments constructed for roads, railway lines, canals etc. and the slopes of earth dams constructed for storing water are examples of man-made slopes.

人工边坡按照要求由人类筑成,通过非自然过程形成。路堑两侧,道路、铁道、运河等的筑堤边坡,以及蓄水土坝边坡,均是人工边坡的实例。

Man-made slopes exist all over the world. In mountainous areas, natural slopes are adapted for construction of roads and other infrastructure, and design of man-made slopes becomes an inevitable and important problem. Because of the wide variety of slope failure, complex genesis of slope failure and limitation of exploration data, usually the original design scheme cannot agree with the practical circumstances very well, thus the observational method in design becomes very important.

世界各地都有人工边坡。在山区,自然边坡适合于建设道路和其他基础设施,设计人工边坡成为一个不可避免的重要问题。由于边坡破损的种类繁多,且成因复杂,以及勘探数据的局限,起初的设计方案通常不能很好地切合实际环境,因此设计中的观测方法变得十分重要。

A man-made slope is a protection facility that has been artificiallyexcavated, then filled up to block the earth from moving down. These may be considered in three main categories:

人工边坡是一种防护设施,人工挖掘,再填满、填实,以阻挡泥土下移。这些可分为三大类:

1. 路堑边坡(Cut slopes)

Shallow and deep cuts are of major interest in many civil and mining engineering operations.The aim is to design a slope with such a height and inclination as to be stable within a reasonable life span and with as much economy as possible. Such design is influenced by geological conditions, material properties, seepage pressures, the possibility of flooding and erosion, the method of construction as well as the purpose of a particular cutting. In mining operations excavations may be carried out in several steps or benches and the stability of individual benches must be ensured as well as of the entire cut.

浅、深路堑是许多土木和采矿工程作业的主要关注点。其目的是设计的边坡高度和倾角在合理的寿命期内都保持稳定且尽可能节约。这种设计受地质状况、物资性质、渗透压、洪水和侵蚀的可能性、施工方法以及特定开挖作业目的的影响。在采矿作业中,开挖可按几步或几级台阶进行,各台阶以及整个开采面的稳定性也必须予以确保。

Steep cuts may sometimes be necessary in many engineering applications so that preventive and protective measures are parts of the initial design. In some situations the stability at the end of construction of a cutting may be critical. On the one hand, many cut slopes are stable in the short-term but may fail without much warning many years later. The most well-known example is that offailures of cut slopes in London clay (Figure 6.2). Making cut slopes so flat that they are stable for an indefinite period of time would often be uneconomical and sometimes impractical. On the other hand, slopes which are too steep may remain stable for only a short time and pose real danger to life and property. Frequent failures would also involve tremendous inconvenience and the expense of repairs, maintenance and stabilization measures.

在许多工程应用中,陡峭路堑有时是必要的,因此预防和保护措施是初步设计的一部分。在某些情况下,路堑施工结束时的稳定性会很关键。一方面,许多路堑边坡在短期内稳定,但多年后会在没有多少预警的情况下破损。最著名的例子是伦敦黏土路堑边坡的破损(图 6.2)。使路堑边坡如此平坦,以致其在未限定期内保持稳定,往往会不节俭,有时不切实际。另一方面,过于陡峭的边坡仅可在短期内保持稳定,长期来看会对生命和财产构成真正危险。边坡频繁损坏不仅会带来极大的不便,并且会涉及修理、维护和稳定措施的费用。

2. 包括土坝的路堤(Embankments including earth dams)

Fill slopes involving compacted soils include railway and highway embankments, earth dams and levees. The engineering properties of materials used in these structures are controlled by the method of construction and the degree of compaction. The analysis of embankments does not involve the same difficulties and uncertainties as does the stability

图 6.2 伦敦黏土路堑边坡损坏情景
(Figure 6.2 Failures of cut slopes in London clay)

of natural slopes and cuts. However, independent analyses are required for the following critical conditions: (i) end of construction, (ii) long-term condition, (iii) rapid drawdown (for water-retaining structures like earth dams), and (iv) seismic disturbance. In recent years the advantages of an observational approach have been demonstrated and it is usual to monitor the performance of embankments and earth dams during and after construction. The construction of test sections of embankments is particularly useful for large projects.

涉及压实土的填方边坡包括铁路和公路路堤、土坝和堤坝。这些结构中所用材料的工程性质由施工方法和压实度控制。路堤分析不像天然边坡和路堑的稳定性那样的困难和不确定。然而,以下关键条件要求独立分析:①施工结束;②长期条件;③快速下降(对于像土坝这样的挡水结构)和④地震性干扰。近年来,观测方法的优点已经得到证明,通常在施工期间和以后监测路堤和土坝的性能。路堤试验段的施工对大型项目特别有用。

It is often necessary to consider the stability of an embankment-foundation system rather than that of an embankment alone.In major projects it is often economically feasible to conduct comprehensive and detailed investigations of foundation conditions. However, in many cases embankments have to be built on weak foundations so that failures by sinking, spreading and piping effect can occur irrespective of the stability of embankment slopes.

常常有必要考虑路堤基础系统的稳定性,而不是单单考虑路堤。在重大项目中,对地基状况进行全面细致的调查往往在经济上可行。然而,在许多情况下,路堤必须建在薄弱基础上,于是,不管路堤边坡稳定性如何,沉降、铺展和管涌带来的损坏都能发生。

The most recent example of a major slope failure due to piping and internal erosion is that of the Teton Dam, Idaho, USA on June 5, 1976 (Figure 6.3). Internal erosion and piping effect occurred in the core of the dam deep in the right foundation key trench. Soil particles moved through channels along the interface of the dam with the highly pervious abutment rock and talus. The volcanic rocks at the site were intensely fissured and water was able to move rapidly during reservoir filling. The wind-deposited clayed silts of very low permeability used for the core and key trench are highly erodible.

1976 年 6 月 5 日,美国爱达荷州的蒂顿大坝,就是由于管涌和内部侵蚀而导致重大边坡破损的最新示例(图 6.3)。内部侵蚀和管涌发生在坝体核心部位,深埋在坝体右侧基础的关键沟槽内。土颗粒沿坝肩高渗透岩与滑石的界面通过渠道移动。该区火山岩裂隙发育,储层充填期间水能够快速运移。用于核心和关键沟渠的渗透性极低的风积黏性粉土极易被侵蚀。

图 6.3 美国爱达荷州的蒂顿大坝

(Figure 6.3 Teton Dam, Idaho, USA)

An independent review panel concluded that the use of this material adjacent to intensely jointed rock was a major factor in the failure. They also felt that the geometry of the key trenches favored arching, reduction of normal stress and consequent development of cracks in the erodible fill. Cracking by hydraulic fracturing was considered to be another possibility since calculations showed that water pressure at the base of the key trench could have exceeded the sum of lateral stresses in the impervious fill and its tensile strength. Whatever the initial cause of cracking, it led to the opening of channels through the erodible fill. Once piping effect began it progressed rapidly through the main body of the dam leading to complete failure.

独立的评审小组得出结论:使用这种材料邻近节理剧烈活动的岩石是导致破损的重大因素。还认为:关键沟渠的几何形状促成拱起、法向应力降低以及随后可蚀填料中裂缝的发展。认为水力压裂裂缝是另一种可能性,因为计算表明,关键沟槽基部的水压可能已超过防渗填料的侧向应力和抗拉强度之和。无论开裂最初原因是什么,都导致通过可腐蚀填料的渠道打开。管涌一旦开始,就穿过大坝主体迅速传开,导致其完全破损。

3. 矸石堆(Spoil or waste heaps)

The stability of spoil heaps consisting of mining and industrial waste is being recognized as a problem of major importance in view of the many recent disasters which have been a consequence of failures of spoil heaps, the growing magnitude of wastes requiring to be disposed in this manner, and the scarcity of adequate sites for waste dumps (Figure 6.4). Until recently (about 10 years ago) spoil heaps had little or no compaction control and in many cases, compaction was not even considered. There are some instances in which particulate wastes may be uniform in composition and engineering properties. However, in general, the problems are somewhat different from those concerning embankments due to differences in methods of construction, uncertainties in geotechnical characteristics and foundation conditions which are often unfavorable. The solution of these problems is greatly complicated where there is inadequate control on composition, location and compaction of the refuse materials.

由采矿废料和工业废料组成的矸石堆,其稳定性被认为是一个非常重要的问题,这是由于最近的许多灾害一直都是矸石堆损坏的结果,要求以这种方式处置的矸石体量越来越大,以及缺乏适当的倾卸场(图 6.4)。直到最近(大约 10 年前),矸石堆几乎或根本没有压实控制,在许多情况下甚至不考虑将其压实。有几个实例,颗粒废物的成分和工程性质会始终如一。然而,总的来说,由于施工方法的差别、岩土特征的不确定性和地基状况的不利因素,这些问题与路堤涉及的有所不同。这些问题解决起来非常复杂,对矸石的成分、位置和压实的控制不当。

图 6.4 矸石堆

(Figure 6.4 Waste heaps)

Where refuse materials are placed in a loose state, shear failure is often followed by liquefaction (complete loss of strength) with catastrophic consequences. Fortunately such occurrences are now rare because of increasing awareness leading to regulationand control. Initiation of instability is often a result of inadequate drainage in wet, saturated dumps. Failure may also occur due to overtopping caused by inadequate spillway capacity in tailings dams.

矸石材料以松散状态存放之处,剪切受损常紧随着液化(强度完全丧失),带来灾难性后果。幸好,由于监管控制意识的提高,此类事件现在很少。不稳定性的始发,往往是由于潮湿、饱和的倾倒区排水不足。由于尾矿坝溢洪道能力不足,也可发生溢流破损。

6.3 边坡防护(Slope Protection)

6.3.1 边坡稳定性(Slope stability)

Landslides, slips, slumps, mudflows, rockfalls — these are just some of the terms which are used to describe movements of soils and rocks under the influence of gravity. These movements can at best be merely inconvenient, but from time to time they become seriously damaging or even disastrous in their proportions and effects. We are normally more aware of hazards arising from the earth's surface processes in terms of flooding and short-term climatic effects, but in other parts of the world, slope instability too, is widely recognized as an ever-present danger. Landslides and other gravity-stimulated mass movements are important and costly problem, and they are a continual source of concern for geotechnical engineers and engineering geologists throughout the world, particularly in geologically "active" regions.

滑坡、滑塌、塌陷、泥石流、落石,只是描述在重力影响下土壤和岩石运动的几个术语。这些运动最多只是造成麻烦,但按其程度和效应,不时会变成严重损害,甚至灾难。正常情况下更清楚灾害出自地表过程按术语讲是洪涝灾害和短期气候效应,但在世界其他地区,边坡不稳定也普遍认为是一种一直存在的危险。滑坡和其他重力引起的团块运动是一个重要且代价高昂的问题,也是全世界岩土工程师和工程地质学家持续关注的问题,尤其在地质"活跃"区。

Slope stability is the potential of soil-covered slopes to withstand and undergo movement. Stability is determined by the balance of shear stress and shear strength. A previously stable slope may be initially affected by preparatory factors, making the slope conditionally unstable. Triggering factors of a slope failure can be climatic events can then make a slope actively unstable, leading to mass movements. Mass movements can be caused by increase in shear stress, such as loading, lateral pressure, and transient forces. Alternatively, shear strength may be decreased by weathering, changes in pore water pressure, and organic material.

边坡稳定性是土壤覆盖的斜坡承受和经受运动的潜力。稳定性是由剪切应力和剪切强度的平衡确定的。原先稳定的边坡,最初可受到引导性因素的影响,使边坡在一定条件下不稳定。诱发边坡破损的因素是气候事件,继而能促使边坡主动失稳,导致团块运动。这种运动由剪切应力,例如荷载、侧压力和瞬态力的增加引起。或者说,抗剪强度可能会因风化、孔隙水压力变化和有机材料而降低。

The field of slope stability encompasses static and dynamic stability of slopes of earth and rock-fill dams, slopes of other types of embankments, excavated slopes, and natural slopes in soil and soft rock. Slope stability investigation, analysis (including modeling), and design mitigation is typically completed by geologists, engineering geologists, or geotechnical engineers. Geologists and engineering geologists can also use their knowledge of earth process and their ability to interpret surface geomorphology to determine relative slope stability based simply on-site observations.

边坡稳定性的研究领域包括土石坝、其他类型堤防、开挖、天然土质和软岩两类边坡的静、动稳定性。边坡稳定性调查、分析(包括建模)和设计缓解通常由地质学家、工程地质学家或岩土工程师完成。地质学家和工程地质学家能凭借所具备的知识和能力来解释地表地貌,仅仅靠现场观察就确定出边坡的相对稳定性。

Factors affecting the stability of slope, inclnding:

影响边坡稳定性的因素有:

(1) Topography and its surrounding physical conditions.

(1) 地形及其周围的物理条件。

(2) Geological conditions such as the nature and depth of its subsoil, degree of decomposition, or location of fracture etc.

(2) 地质状况,例如其底土的自然性质和深度、分解程度或断裂部位。

(3) Shear strength of the slope-forming materials.

(3) 边坡形成材料的抗剪强度。

(4) Surface and ground water condition.

(4) 地表水和地下水状况。

(5) External loading and surcharges, such as from traffic, nearby structures, possible vibration.

(5) 外部附加费用和额外负担,例如出自交通、附近结构以及可能的振动。

6.3.2 实例(Examples)

A fall of material, soil or rock, is characteristic of extremely steep slopes. The material which moves can break away from the parent rock by an initial sliding movement: some shear surfaces may develop in response to gravity stresses and in moving, the material is projected out from the face of the slope. Alternatively, due to undermining at a low level in the slope, an overhang may form. Causes for the undermining may include wave action, river or stream erosion, erosion of an underlying bed by seepage, weathering or careless excavation; they therefore include both internal and external agencies. Then, either because the rock is jointed, or because it has insufficient strength *en masse*, there comes a point at which the undermining causes a fall to occur. Progressive weakening, perhaps by weathering of a susceptible unit in a cliff, can also allow joint-bounded blocks to rotate until they pass through a position of equilibrium and overtopple. As the blocks rotate, they throw more stress on the outside edge and it is self-evident that this must accelerate the onset of toppling.

材料、土壤或岩石的坠落,是非常陡峭斜坡的特征。移动的材料能靠初始滑动运动脱离母岩一定距离:一些剪切面可在响应重力应力中培育,在移动过程中,材料从边坡表面突出;或者,由于在斜坡的低水平处进行破坏,因此可形成一种悬挑。造成破坏的原因可包括波浪作用、河流溪流侵蚀、下卧层渗透侵蚀、风化或草率开挖;因此,还包括内外部能动作用。然后,或是因岩石属于节理式,或因岩石的整体强度不足,在这种情况下,就出现一个部位,破坏作用在此导致一次坠落。或许是受悬崖中敏感单元的风化作用，逐渐弱化也能使节理束缚的块体旋转，直至通过平衡位置并倒塌翻滚。随着块体旋转，向外缘投出更多应力，这必然会加速倾倒的发生。

The effect of water pressures in a joint-bounded rock mass should not be underestimated. Where the natural outlets become blocked, by ice formation as a typical example, extremely large thrusts can be developed. Ice wedging itself, if the water in the joints freezes, can also generate significant forces. These may be sufficient to rupture unjointed rock. Failure may follow immediately, or wait until the thaw. Ice formation is not absolutely necessary to the mechanism: unfrozen water exerts high thrusts if the joints are full. An example of the combined effects of high joint water thrusts and scour can be seen in the American Falls at Niagara, Figure 6.5, where massive joint-bounded blocks of Lockport Dolomite lie piled against the slope face. These have been dislodged partly by the undermining of the sandstones and shales beneath the dolomite, but a consideration of the forces involved shows that water thrusts have an important role in destabilizing the blocks.

不宜低估节理束缚的岩体中水压效应。典型例子是，在天然出口被冰层堵塞之处，能形成巨大的推力。节理中的水冻结时，冰楔作用本身也能产生甚大的力，可以大到足以将无节理岩石破裂。破损可能紧随而来，或者待到解冻。冰层对这一机制并非绝对必要：节理到填满时，未冻结的水产生高推力。从靠近美国尼亚加拉瀑布的岩块滑坡(图 6.5)中，能看到高节理水推力和冲刷的联合效应的一个实例，在此，洛克波特白云岩节理束缚的巨大岩块堆积在坡面上。由于白云岩下方砂岩和页岩的破坏，这些岩石部分移动，但从所涉及的力考虑，表明水冲推力在块体失稳上起重要作用。

图 6.5 靠近尼亚加拉瀑布的洛克波特白云岩块滑坡

(Figure 6.5 Destabilizing the blocks of Lockport dolomite, by Niagara Falls)

Even in arid regions, a form of wedging can occur. In response to daily, and to annual, temperature changes, joints open and close. When open, small pieces of debris fall down the joint, preventing proper closure when the temperature drops. The effect of this would be progressive. Seismic shocks, too, can dislodge debris from steep slopes.

即使在干旱地区，也能出现一种楔入作用形式。根据每日每年的温度变化，节理系时开时闭。打开时,小碎片从节理掉落,阻止在温度下降时其正常闭合。这种效应会是渐进式的。地震性冲击也能把陡坡上碎片扯下。

As seen in Figure 6.6, earthen slopes can develop a cut-spherical weakness area. The probability of this happening can be calculated in advance using a simple 2D circular analysis package. A primary difficulty with analysis is locating the most-probable slip plane for any given situation. Many landslides have only been analyzed after the fact. More recently slope stability radar technology has been employed, particularly in the mining industry, to gather real time data and assist in pro-actively determining the likelihood of slope failure.

如图 6.6 所示,土坡可以培育一种割球面状软弱区。此情发生概率能用简易的二维循环分析包提前算出。分析中的一个主要难点是为给定情势定位最可能的滑动面。许多滑坡仅在事后进行分析。最近,边坡稳定性雷达技术已经被采用,特别是在采矿业,收集实时数据并帮助确定边坡破损的可能性。

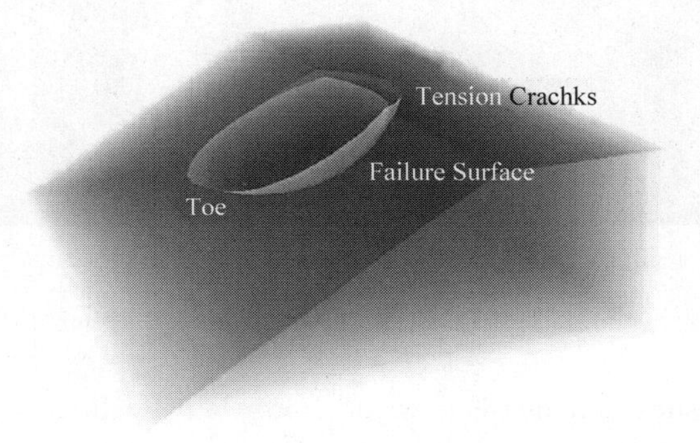

图 6.6 简单斜坡滑动段

(Figure 6.6 Simple slope slip section)

As seen in Figure 6.7, Real life failures in naturally deposited mixed soils are not necessarily circular, but prior to computers, it was far easier to analyze such a simplified geometry. Nevertheless, failures in "pure" clay can be quite close to circular. Such slips often occur after a period of heavy rain, when the pore water pressure at the slip surface increases, reducing the effective normal stress and thus diminishing the restraining friction along the slip line. This is combined with increased soil weight due to the added groundwater. A "shrinkage" crack at the top of the slip may also fill with rain water, pushing the slip forward. At the other extreme, slab-shaped slips on hillsides can remove a layer of soil from the top of the underlying bedrock. Again, this is usually initiated by heavy rain, sometimes combined with increased loading from new buildings or removal of support at the toe. Stability can thus be significantly improved by installing drainage paths to reduce the destabilizing forces. Once the slip has occurred, however, a weakness along the slip circle remains, which may then recur at the next monsoon.

如图 6.7 所示,在自然沉积的混合土壤中,现实的破损未必是圆的,但在计算机出现之前,分析这种简化的几何结构已经相当容易。然而,"纯"黏土的破损能非常接近于圆形。这种滑移往往发生在暴雨期后，此时滑移面孔隙水压力增大，有效正应力减小，沿滑移线的约束摩擦力减小。这与固料增加了地下水土壤即增重相结合。滑移顶部的“收缩”裂缝也可充满雨水，推动滑移向前。在另一个极端是，山坡上的板状滑动可能将底层基岩顶部的一层土壤移走。再者，这通常是由大雨引发的，有时与新建筑所增荷载或趾状支撑移除相结合。因此，通过安装排水道来减少矢稳力，能显著提高稳定性。然而，一旦发生滑移，沿滑移圈的一个弱点仍然存在，并可在下次季风时重现。

图 6.7 现实生活中的边坡滑坡

(Figure 6.7 Real life landslide on a slope)

Slope stability issues can be seen with almost any walk down a ravine in an urban setting. An example is shown in Figure 6.8, where a river is eroding the toe of a slope, and there is a swimming pool near the top of the slope. If the toe is eroded too far, or the swimming pool begins to leak, the forces driving a slope failure will exceed those resisting failure, and a landslide will develop, possibly quite suddenly.

在城市环境中，沿峡谷而下的几乎所有步道都存在斜坡稳定性问题。图 6.8 所示的例子中，一条河流正侵蚀着坡脚，坡顶附近有个游泳池。当坡趾侵蚀得太严重，或者游泳池开始漏水时，驱动斜坡破损的力会超过那些抵挡破损的力，滑坡就会发育，还可能十分突然。

图 6.8 伴有侵蚀河流和游泳池的斜坡

(Figure 6.8 Slope with eroding river and swimming pool)

6.3.3 边坡失稳模式(Identification of modes of slope instability)

Different types of slope failure are associated with different geological structures and it is important that the slope designer be able to recognize potential stability problems during the early stages of a project.

不同类型的边坡破损与不同地质结构关联,重要的是,边坡设计者能够识别项目早期的潜在稳定性问题。

A fairly common engineering failure is slipping of an embankment or cutting, and considerable research has been carried out into the causes of such failures. Water is frequently the cause of earth slips, either by eroding a sand stratum, lubricating a shale or increasing the moisture content of a clay, and hence decreasing the shear strength. When a slip in a clay soil occurs it is frequently found to be along a circular arc, and therefore this shape is assumed when studying the stability of a slope. This circular arc may cut the face of the slope, pass through the toe or be deep-seated and cause heave at the base (see Figure 6.9).

路堤或路堑滑动是相当常见的工程损坏形式,对这些破损的成因已进行了大量研究。水常常是造成土方滑动的原因,或由于腐蚀了砂土层、润滑了页岩或由于增加了黏土的水分而降低了其抗剪强度。研究发现,当黏土发生滑动时,发现常常是沿着圆弧,因此,在研究斜坡稳定性时就假定是这种形状。这种圆弧可切开坡面,穿过坡或达到深座,导致底部隆起(图 6.9)。

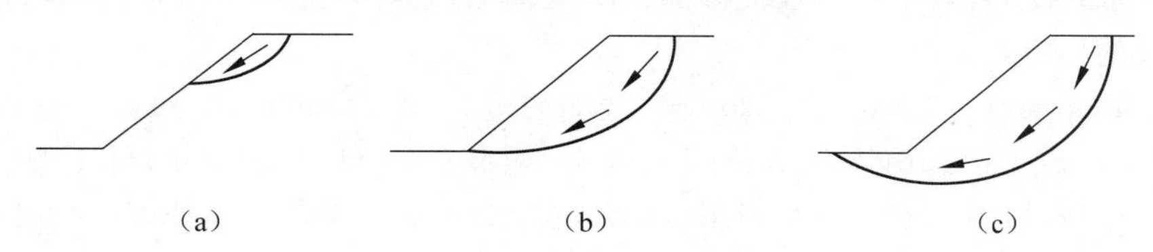

图 6.9 边坡破损模式

(a) 斜坡破损;(b) 坡趾破损;(c) 深座破损

(Figure 6.9 Slope failure mode)

(a) Slope failure; (b) Toe failure; (c) Deep-seated failure

The cause of slope failure in a cutting will be quite different from that in an embankment. A cutting is an unloading case where soil is removed, hence causing relief of stress in the soil. The soil resistance dissipates with time and a part of the engineer's problem will be to predict the soil properties during the design life of a cutting.

路堑边坡与路堤边坡失稳的原因有很大不同。路堑是因卸载,土壤移除使土壤压力得到缓解。土壤阻力随着时间消散,工程师的部分任务将是在设计路垫寿命期间预测土壤的性质。

Embankments and spoil heaps, on the other hand, are loading cases and the construction period is the most critical period, owing to the buildup of pore pressures during construction, with the consequent reduction in effective stress. In time, these excess pore pressures dissipate and the shear resistance of the embankment increases, although consolidation may now become the major problem.

另一方面，堤坝和矸石堆都属于加荷装载的情况，施工期最关键，因为期间孔隙压力聚集并导致有效应力随之降低。随着时间推移，这些超额的孔隙压力逐渐消散，路堤剪切阻力增强，固结可能变成目前的主要问题。

In both cases the study of the variation of pore-water pressure within the soil mass is of paramount importance, since only in this way can reasonable values of the parameters c and φ be determined. Water seepage will set up seepage pressures which may induce failure of the slope. Such water movement frequently occurs in the vicinity of major earthworks such as an earth dam or cuttings below the natural water table. Seepage pressures can be monitored with piezometers on site and it may be necessary to provide suitable drainage to control the flow of water.

在两种情况下，研究土体内孔隙水压变化至关重要，因为只有这样方能确定参数 c 和 φ 的合理值。水的渗流会产生渗流压，可引起边坡受损。水的这种运动经常发生在大型土方工程附近，例如土坝或天然水位以下的岩屑。渗透压可在现场用压力计监测，这会有必要提供适当的排水系统以控制水的流动。

From the preceding discussion, it will be seen that control of water must be considered. Suitably designed drainage should minimize any seepage pressures which may be set up and will also reduce pore-water pressures, thus increasing effective stress, and therefore, the stability of the slope.

从前面讨论中看出，(在设计斜坡或试图稳定现有斜坡时)必须考虑对水进行控制。设计适当的排水系统，宜可形成的任何渗漏压降至最低，且降低孔隙水压，增加有效应力并因此增加斜坡的稳定性。

In embankment slopes, horizontal layers of coarse material may be included to facilitate drainage and provision must be made to dispose of the water from these layers.

在路堤边坡，可纳入水平的粗料层，以方便排水，并必须采取措施处置这些层中的水。

In cuttings, surface drainage will prevent softening of the upper layers of soil, but will do little to increase overall stability. Installation of deep-seated drainage in a cutting can be very expensive and some method of loading, or unloading, the slope may provide a better solution.

在路堑中，表面排水会阻止上层土软化，但对增加整体稳定性作用不大。在路堑中安装深层排水系统会非常昂贵，而某些装载或卸载的方法在坡度上可提供更好的解决方案。

For natural hillslopes, the slip surface is generally along a plane parallel to the ground surface and at a fairly shallow depth. In this situation, surface drainage, provided it reaches beyond the failure plane, may well be sufficient.

对于天然边坡而言，滑面一般是沿着与地表平行的平面，深度甚浅。在此情况下，倘若地表排水能超出破损面，就算足够好了。

6.3.4 分析方法和加固设计(Analysis method and reinforcement design)

If the forces available to resist movement are greater than the forces driving movement, the slope is considered stable. A factor of safety is calculated by dividing the forces resisting movement by the forces driving movement. In earthquake-prone areas, the analysis is typically run for static conditions and pseudo-static conditions, where the seismic forces from an earthquake are assumed to add static loads to the analysis.

当可用于抵抗运动的力大于驱动运动的力时，认为边坡稳定。安全因数的计算通过抵抗运动的力除以驱动运动的力。在地震多发区，该分析通常针对静态和伪静态条件，在此条件下，地震产生的地震力被假定为静态载荷。

When it has been established that a slope is potentially unstable, reinforcement may be an effective method of improving the factor of safety. Methods of reinforcement include the installation of tensioned anchors or fully grouted, untensioned dowels, or the construction of a toe buttress. Factors that will influence the selection of an appropriate system for the site include the site geology, the required capacity of the reinforcement force, drilling equipment availability and access, and time required for construction.

当已确认边坡潜在不稳定时，加固是可提高安全因数的有效方法，包括安装拉紧式锚或完全灌浆，未张紧式销钉或建造趾式扶壁。影响选择合适系统的因素包括：场地地质，所需加固能力，钻孔设备的可用性和通道，以及施工所需时间。

If rock anchors are to be installed, it is necessary to decide if they should be anchored at the distal end and tensioned, or fully grouted and untensioned. Untensioned dowels are less costly to install, but they will provide less reinforcement than tensioned anchors of the same dimensions, and their capacity cannot be tested. One technical factor influencing the selection is that if a slope has relaxed and loss of interlock has occurred on the sliding plane, then it is advisable to install tensioned anchors to apply normal and shear forces on the sliding plane. However, if the reinforcement can be installed before excavation takes place, then fully grouted dowels are effective in reinforcing the slope by preventing relaxation on potential sliding surfaces Untensioned dowels can also be used where the rock is randomly jointed and it is necessary to reinforce the overall slope, rather than a particular plane.

必须安装岩石锚时，有必要决定宜在远端锚定并拉紧还是完全灌浆且不拉紧。未拉紧销钉安装成本较低，但提供的加固少于同等尺寸的张拉锚，能力也无法测试。影响选择的一个技术因素是，当边坡已经松弛并且在滑动面上发生互锁损失时，建议安装拉紧式锚以便在滑动面上施加法向和剪切力。然而，当开挖前能安装加固件时，只完全灌浆的销钉对防止潜在滑动面松弛，来加固边坡有效。未拉紧销钉也能用在岩石随机形成节理处，这对加强整个斜坡总体，而非特殊平面很有必要。

6.4 专业词汇(Specialized Vocabulary)

slope engineering 边坡工程
natural slope 自然边坡
man-made slope 人工边坡
Slope protection 边坡防护
geological adj. 地质的，地质学的

ore n. 矿;矿石
hilly adj. 丘陵的;多坡的
mining n. 矿业;采矿
geomorphologist n. 地貌学家
gravitational adj. 重力的,引力的
shearing resistance 剪切阻力
seismic adj. 地震的;地震引起的
deforestation n. 毁林,森林采伐
drainage n. 排水(系统)
urbanization n. 城市化
pore water pressure 孔隙水压力
shear strength 剪切强度
weathering n. 风化作用;泄水斜度
overconsolidated adj. 超固结的
clay n. 黏土;泥土
shale n. 页岩
diagenetic adj. 成岩作用的
valley joints 山谷节理系
erosion n. 侵蚀,腐蚀
solifluction n. 泥流;泥流作用
excavate v. 挖掘;开凿
cut slope 路堑边坡
embankment n. 路堤;堤防
sinking n. 沉降
piping n. 管道
abutment n. 桥墩;支撑(点)
tensile strength 抗拉强度
spoil or waste heap 矸石堆
liquefaction n. 液化;溶解
spillway n. 溢洪道;泄洪道
slope stability 边坡稳定性
reinforcement n. 加固;加强
buttress n. 扶壁;扶壁状凸起

习题(Exercises)

1. Translate the following sentences into Chinese.

(1) Geology, nature of slope materials and topography are always important factors in the case of natural slopes and often in the case of man-made slopes.

(2) Natural slopes are those that exist in nature and are formed by natural causes.

(3) Deformations often occur along major natural discontinuities, ancient slip surfaces and tectonic shear zones within a slope.

(4) Fill slopes involving compacted soils include railway and highway embankments, earth dams and levees.

(5) Slope stability is the potential of soil covered slopes to withstand and undergo movement.

(6) Slope stability issues can be seen with almost any walk down a ravine in an urban setting.

2. Translate the following sentences into English.

(1) 边坡是人类生存和工程活动最基本的自然地质环境之一。

(2) 边坡材料的应力史具有十分重要的意义。

(3) 人工边坡是人类根据需要建造的。

(4) 边坡由非自然过程形成。

(5) 通常有必要考虑路堤基础系统的稳定性,而不是仅仅考虑路堤的稳定性。

(6) 如果其抵抗运动的力大于驱动运动的力,则认为边坡是稳定的。

城区地下空间
(Urban Underground Space)

7.1 地铁工程(Subway Engineering)

7.1.1 岩土工程调查(Geotechnical investigations)

Subsurface (geotechnical) investigations will further evaluate geology, groundwater, seismic, and environmental conditions along the alignment. These investigations would be spaced along the alignment to evaluate soil, rock, groundwater, seismic and geo-environmental conditions, particularly to note locations where hydrocarbon or other contaminant deposits may be encountered. The results of these investigations will influence final design and construction methods for stations, tunnels, other underground structures, and foundations.

地下(岩土)调查会进一步评价沿线的地质、地下水、地震和环境条件。这些调查将沿线进行,以评估土壤、岩石、地下水、地震及地质环境条件,对于所遇到含烃或污染性沉积物的部位予以特别标明。调查结果将影响到车站、隧道、其他地下建筑以及基础的最终设计和施工方法。

7.1.2 车站施工(Station construction)

The construction of underground stations would employ the cut-and-cover construction technique. This technique generally begins by opening the ground surface to an adequate depth to permit support of existing utility lines and to install soldier piles, or other earth-retaining structures. The surface opening is then covered with a temporary street decking so traffic and pedestrian movement can continue overhead while excavation proceeds beneath the decking. The temporary excavation will be retained by an approved excavation support system, known as a shoring system. Adjacent building foundations will also be supported as necessary. A concrete station box structure is then built within the excavated space, backfilled up to street level, and the surface is restored.

地下车站施工会采用“明挖法”。这项技术是从地面向下开挖到适当深度,以支撑现有公共设施管线，并安装支护桩或其他挡土墙结构，然后在地面开口罩设临时性街道甲板式铺板，以使交通和行人继续架空运行，而甲板下方开挖同时进行。临时性开挖将按经核准开挖支护体系(称为支撑系统)予以挡护，临近建筑物地基于必要时也将予以支护。最后在所开挖空间内建造箱型结构混凝土站，回填至路面高度并恢复其表面。

7.1.3 隧道施工(Tunnel construction)

Tunneling is expected to be performed with pressurized-face tunnel boring machines (TBMs). The TBM type used for different reaches of the tunnels will be subject to varying, site-specific requirements, including geologic conditions. For instance, where hydrocarbons and/or gases are expected to be encountered it is likely that a slurry-face TBM will be required. Where there is less contamination, it is expected that either a slurry-face or earth-pressure balance (EPB) TBM will be used. The distinction between these machine types is presented later in this appendix.

预期将用加压面隧道掘进机(TBM)进行隧道掘进作业。用于不同隧道段的TBM类型将视现场特定要求(包括地质状况)而定。例如，当遇有碳氢化合物和/或天然气时，很可能需要泥水平衡掘进机。污染较少处，预计将使用浆面平衡或土压平衡(EPB)TBM。这些机器类型之间的区别将在附录中介绍。

7.1.4 土压平衡掘进机(The earth pressure balance (EPB) TBMs)

The EPB TBMs rely on balancing the thrust pressure of the machine against the soil and water pressures from the ground being excavated.The EPB TBMs are generally well-suited for mining in soft ground, as expected with the proposed project. These TBMs can also mine through variable soils, and groundwater. The excavation method for an EPB TBM is based on the principle that tunnel face support is provided by the excavated soil itself(Figure 7.1).

土压平衡掘进机依靠机器推力去平衡挖掘中地面的水土压力。正如拟建项目预期，这种掘进机一般非常适合在软地层开挖。也能贯通各种土体和地下水。其开挖方法基于所开挖土体本身提供的隧道工作面支撑的原理(图7.1)。

7.1.5 泥水平衡掘进机(Slurry-face balance TBMs)

Slurry-face TBMs will likely be required for tunneling in gassy zones, where the addition of the slurry and the closed spoil removal system provides more protection against gas intrusion into the tunnel environment (Figure 7.2). Where lower gas concentrations are expected, EPB TBMs may be suitable. With Slurry-face TBMs, bentonite (clay) slurry is added in a pressurized environment at the tunnel excavation face. This combination of pressure and slurry stabilizes and supports the soils during excavation. Depending on the ground encountered, conditioners may be added to the slurry. Excavated soil is mixed with the slurry fluid and then pumped out of the tunnel to an above-ground

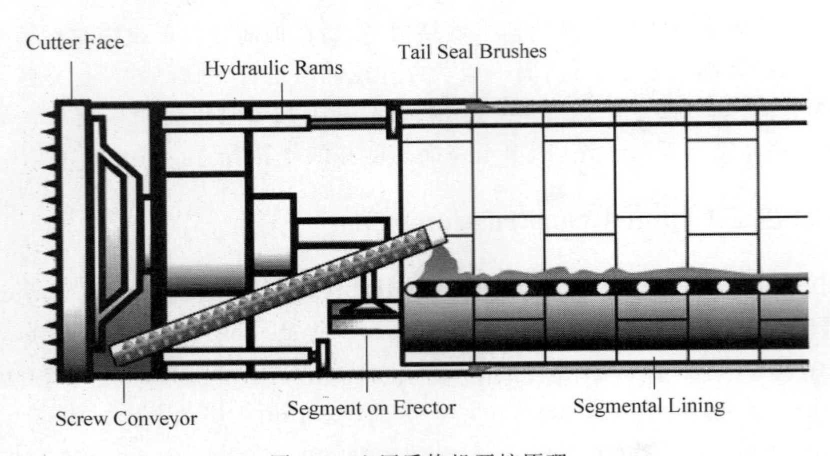

图 7.1　土压盾构机开挖原理

(Figure 7.1　EPB TBM excavation principle)

separation plant through large (approximately 18m diameter) pipelines with in-line booster pumps, the excavated materials are treated at a separation plant, where they are separated from the slurry mixture. This also allows safe dispersion of any potentially gaseous components without endangering tunnel personnel.

在天然气地带隧道掘进很可能需要用泥水平衡掘进机，在此区域，添加泥浆和封闭式除渣系统可提供更多保护，防止天然气体侵入隧道环境(图 7.2)。在天然气浓度预计较低处，土压平衡隧道掘进机非常合适。使用时，在隧道开挖面加压环境中添加膨润土(黏土)浆。压力和泥浆的这种结合在挖掘期间稳定并支撑了土体。根据所遇地层，可将调节剂加进浆料。所挖土体也与泥浆混合，再经由配有直列式增压泵的大型管线(直径约 18m)从隧道中泵至地上分离厂。挖出的材料在分离厂进行处理，土从泥浆混合物中分离出来。这还使得任何潜在气体成分安全消散，而不危及隧道人员。

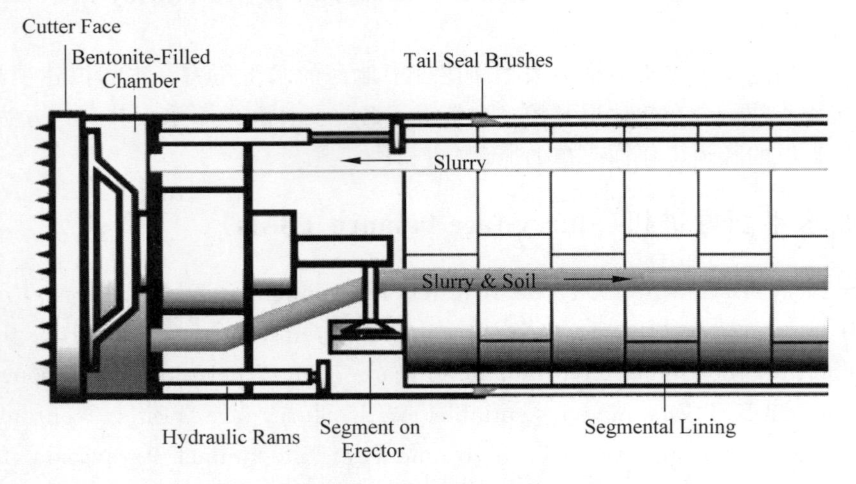

图 7.2　泥浆 TBM 示意图

(Figure 7.2　Schematic diagram of Slurry-face TBM)

7.2 综合管廊(Integrated Pipe Gallery)

7.2.1 地下管廊的建设意义(The construction significance of underground pipe gallery)

Underground pipe gallery system not only solves the problem of urban traffic congestion, but also greatly facilitates the maintenance and overhaul of municipal facilities such as electricity, communication, gas, water supply and drainage. In addition, the system also has a certain earthquake-proof and disaster-reduction role. For example, during the Kobe earthquake in Japan in 1995, a large number of houses collapsed and roads were destroyed, but most of the underground utility corridors were intact, which greatly reduced the difficulty of disaster relief and reconstruction.

地下管廊系统不仅解决了城区交通拥堵问题,还极大便利了电力、通信、燃气、供排水等市政设施的维护和检修。此外,该系统还具有一定的防震减灾作用。例如1995年日本阪神地震期间,大量房屋倒塌、道路被毁,但地下综合管廊大多完好无损,这大大降低了救灾和重建工作的难度。

Underground pipe gallery plays an important role in meeting the basic needs of people's livelihood and improving the comprehensive carrying capacity of cities, reduce the cost of multiple road surface renovations and maintenance cost of process pipelines, and maintain the integrity of road surface and the durability of various pipelines. Due to the compact and reasonable layout of pipelines within the comprehensive pipeline, the space under the road is effectively utilized and urban land is saved. Optimized the landscape of the city due to the reduction of road poles, inspection wells and inspection rooms of various pipelines.

地下管廊对满足民生基本需求和提高城市综合承载能力,降低路面多次翻修的费用和过程管线的维护费用,以及保持路面完整性和各类管线耐久性发挥着重要作用。由于综合管道内管线布置紧凑合理,有效利用了道路下方的空间,节约了城区用地。由于减少了道路的杆柱以及各种管线的检查井、检查室,优化了城市景观。

7.2.2 综合管廊的主体施工(The main construction of comprehensive pipe gallery)

Open excavation: means that the rock (soil) body of the pipe gallery is first removed and then the pipe gallery is built(Figure 7.3). Open excavation has the advantages of simple, fast, economical and safe construction, and it is the preferred excavation technology in the early stage of urban underground tunnel project development. The disadvantage is that it has a greater impact on the surrounding environment. The open excavation method is applicable to the construction of pipe gallery of new roads in new urban areas.

明挖法：是指首先将管廊部位的岩(土)体挖除,然后修建管廊(图 7.3)。明挖法具有施工简单、快捷、经济、安全的优点,是城市地下隧道项目开发初期的首选开挖技术。其缺点则是对周围环境的影响较大。明挖法适用于新城区新道路的管廊施工。

图 7.3 明挖
(Figure 7.3 Open cut)

Shield tunneling method: a fully mechanized construction method. It propels the shield machine in the ground and supports the surrounding rocks around the shield shell and the tube to prevent the collapse into the tunnel. At the same time, in front of the excavation, a mechanized construction method is used to carry out soil excavation with a cutting device, carry out the excavation by the excavated machinery, pressurize the jacking at the back, and assemble precast concrete pipe slices to form a tunnel structure (Figure 7.4). Shield tunneling method is suitable for underground pipe gallery construction in the old city.

盾构法：一种全机械化施工方法。盾构机在地层中掘进,支撑着盾壳和管身四周围岩以防隧道内发生坍塌。同时,在开挖正面使用机械化施工方法,采用切削装置挖掘土体,通过出土机械运出洞外,靠千斤顶在后部加压顶进,并拼装预制混凝土管片,形成隧道结构(图 7.4)。盾构法适合于城市老城区地下管廊施工。

图 7.4 盾构法
(Figure 7.4 Shield-funneling method)

Pipe-jacking method: a method of underground construction when tunnels or underground pipelines cross various obstacles such as railways, roads, rivers or buildings. During construction, hydraulic jacks on the back seat of the foundation pit are used to push the pipe into the soil layer by force-jacking iron and guide rail, while the soil in front of the pipe is removed and carried away. When the first section of the pipe all into the soil, then the second section of the pipe connected in the back to continue to push into, so one pipe jacked into the pipe, the interface formed culvert (Figure 7.5).

顶管法：隧道或地下管线跨越铁路、道路、河流、建筑物等各种障碍物时采用的一种暗挖式施工方法。施工期间，通过传力顶铁和导轨，用基坑后座上的液压千斤顶将管身推入土层，同时挖起并运走管身正面的泥土。当第一节管身全部顶入土中时，就将第二节管身接在后面继续顶进，像这样将一节节管身顶入，接口则形成涵管(图 7.5)。

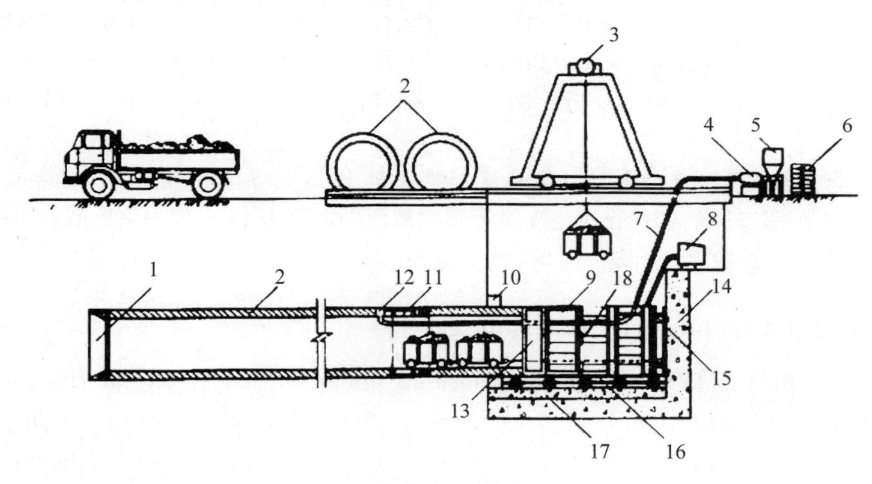

图 7.5 顶管法

1—工具管刃口；2—管身；3—起重行车；4—泥浆泵；5—泥浆搅拌机；6—膨润土；7—灌浆软管；8—液压泵；9—定向顶铁；10—洞口止水圆；11—中继接力环和扁千斤顶；12—泥浆灌入孔；13—环形顶铁；14—顶力支撑墙；15—承压垫水；16—导轨；17—底板；18—后千斤顶。

(Figure 7.5 Pipe-jacking method)

1—tool pipe cutting edge; 2—pipe; 3—crane trolley; 4—slime pump; 5—mud mixer; 6—bentonite; 7—grouting hose; 8—hydraulic pump; 9—directional top iron; 10—water stop circle at portal; 11—relay ring and flat jack; 12—mud-injection hole; 13—circular top iron; 14—jacking support wall; 15—pressure-pad water; 16—guide rail; 17—baseboard; 18—rear jack.

7.2.3 综合管廊分类(The classification of integrated pipe gallery)

1. 干线综合管廊(Trunk line integrated pipe gallery)

The main trunk line corridors are generally located at the bottom of the center of the road and are responsible for providing distribution services to the branch line corridors. The main receiving pipelines are communication, cable TV, electric power, gas and tap water, etc. Some trunk line corridors also include rainwater and sewage system. Its characteristics are: stable and high-flow transportation; a high degree of safety; compact interior; can be directly provided to the stable use of large users; special equipment is generally required; easy to manage and use.

主干线综合管廊一般位于道路中央基底，用于向支线管廊提供配送服务。主要收容管线有通信、有线电视、电力、燃气、自来水等。某些干线管廊还包括雨水、污水系统。其特征是：稳定、大流量输运；高安全性；内部结构紧凑；可直接供给大型用户稳定使用；一般要求有专用设备；易于管理及使用。

2. 支线综合管廊（Branch line comprehensive pipe gallery）

Branch lines are mainly used for distribution and transportation of various supplies from main lines to direct users. Generally located on both sides of the road, it accommodates various pipelines directly serving the areas along the road. The section of branch pipe gallery is usually rectangular, which is usually single or double tank structure.

It is characterized by small effective section, simple structure and convenient construction.

支线综合管廊主要用于将各种补给品从干线分配、输运至各直接用户。它通常位于道路两旁，容纳沿途直接服务区的各种管线。支线管廊的截面以矩形为主，一般为单舱或双舱结构。

其特征为：有效截面小；结构简单，施工方便。

3. 线缆综合管廊（Integrated cable pipe gallery）

The cable duct gallery is generally located under the pavement of the road and is shallow. Rectangular section is more common. General work passage is not required, lighting, ventilation and other equipment are not required in the pipe gallery, only the cover plate or working hand hole can be opened for maintenance.

缆线管廊一般位于道路人行道下方浅部。矩形截面较为常见。一般不需要工作通道，管廊内不要求照明、通风或其他等设备，仅有盖板或工作手孔能在维护时开启。

7.3 地下综合体（Underground Complex）

7.3.1 地下综合体定义（The definition of underground complex）

Underground complexes, also known as underground urban complexes, usually refer to large-scale urban underground spaces that can comprehensively embody urban functions.Generally, the underground complex is set at important urban nodes used to improve ground traffic, expand urban ground space, or protect the environment, etc. In addition, it also resists the hell of war and natural hazard, promote underground public pipe, line facilities and other functions. The underground complex can be combined with newly created city (such as the underground traffic-based large complex located in Des Fontes satellite city, Paris, France), with high-rise buildings (such as the World Trade

Center building in Tokyo, Japan and Rochester Building, New York, USA), or with city squares and streets. The latter, such as the larger Column Alai Square in Paris, France, has 4 floors underground, with a total area of more than 200,000m^2. It can transfer various transportation systems in the downtown area to the ground, and can realize transfer and ground the top is the pedestrian and green area, which is conducive to improving the traffic and environmental conditions in the downtown area (Figure 7.6).

地下综合体,又称城市地下综合体,通常指可综合体现城区功能的大型城市地下空间。一般设置在市区重要节点,用于改善地面交通,扩展城区地面空间、保护环境等。此外,也有抗御战争和自然灾害,提升地下公用管线设施等功用。可与新建城镇结合建设,例如位于法国巴黎德方斯卫星城、以交通为主的地下大型综合体;与高层建筑相结合例如日本东京世界贸易中心大楼、美国纽约罗切斯特大楼等,或与城市广场和街道相结合。例如法国巴黎阿莱广场,地下4层,总面积超过200000m^2,能让市中心多种交通系统都转入地下并实现换乘,大广场顶部则为步行和绿化区,有利于改善市中心的交通和环境状况(图7.6)。

图7.6 法国巴黎阿莱广场

(Figure 7.6 Column Allais, Paris, France)

7.3.2 地下综合体类型(The type of underground complex)

1. 新镇地下综合体(The underground complex of new towns)

In the public activity center of new towns or large residential areas, in conjunction with public buildings on the ground, a part of the functions of transportation and commerce will be placed in the underground complex, which can save land, make the central area walkable, and overcome the effects of bad weather. This underground complex has a compact layout and is easy to use. The ground and underground spaces are integrated into one, which is very popular among residents.

在新镇或大型居住区的公共活动中心,与地面公共建筑相配合,将一部分交通、商业等功能置于地下综合体中,既能节省用地,又使中心区步行化,并克服了恶劣气候的影响。这种地下综合体布置紧凑,使用舒适,地面地下空间融为一体,很受居民欢迎。

2. 与高层建筑群结合的地下综合体(The underground complex combined with high-rise buildings)

The contents and functions of the complex attached to the basement of the high-rise building are mostly related to the nature and function of the high-rise building, which can be regarded as the extension of the ground buildings' function to the underground space.

附建在高层建筑地下室的综合体,其内容与功能大多与该高层建筑的性质和功能有关,能将其视为地面建筑功能向地下空间的延伸。

This high-rise building has such main functions as office and business activities, the content of its underground complex is mostly the facilities that serve the building, such as parking lot, various mechanical and electrical equipment and pipes/lines.

以办公和商务活动为主要功能的高层建筑,其地下综合体的构件大多是为本大厦服务的设施,例如停车场、各种机电设备和管线。

3. 城市地下广场和街道综合体(The underground complex under the city square and street)

It's suitable for building underground complex where in the city's central square, cultural leisure square, shopping center square and traffic square, and traffic or commercial high concentration of streets and street intersections.

在城市的中心广场、文化休闲广场、购物中心广场、交通集散广场,以及交通或商业高度集中的街道和交叉路口处,都适合建造地下综合体。

First of all, in these locations, various urban contradictions, especially traffic contradictions are particularly prominent, therefore, it is also the main location of urban redevelopment.

首先,在这些地点,各种城市矛盾,尤其是交通矛盾特别突出,因而也是城市再开发的主要位置。

Secondly, the underground space of squares and streets is relatively easy to develop, especially squares, the demolition of buildings and underground pipes and lines, and their impact on ground transportation are small.

其次,广场和街道的地下空间相对容易开发,尤其是广场,建筑物和地下管线的拆迁及其对地面交通的影响都小。

7.3.3 地下综合体特征(The characteristics of underground complex)

It can be used as an effective anti-aircraft and anti-explosive facility, forming a constant temperature/humidity, and vibration-proof environment, and can save the floor space occupied by buildings. However, the requirements for geological conditions are high, the construction is difficult, and the investment is high.

能用作一种有效的防空、防爆设施,形成恒温、恒湿、防震、防振的环境,并能节省建筑物所占建筑面积。但对地质状况要求高,施工难,投资也大。

7.3.4 地下综合体实例(Examples of underground complexes)

1. 光谷广场综合体(Optical Valley Plaza complex)

The total construction area of the complex is about 160000m², equivalent of 21 qualified football court; the maximum excavation depth is 34m, which is like digging 11 stories of tall buildings underground. The circular square with a diameter of 200m, is divided into three layers underground, and there is also a mezzanine (Figure 7.7). The complex integrates rail transit, municipal and underground public space, mainly including three major projects:

该综合体总建筑面积约160000m²,相当于21个标准足球场;最大开挖深度34m,好比在地下挖出11层高楼。圆形广场直径200m,地下3层,还有一处夹层(图7.7)。综合体集轨道交通、市政、地下公共空间于一体,主要包括以下3大项目:

Three subway lines: South extension of Line 2, Line 11, Line 9.

Two municipal highway tunnels: LuMo Road Tunnel, LuYu Road Tunnel.

3条地铁线:2号线南延线、11号线、9号线。

2条市政公路隧道:鲁磨路隧道、珞瑜路隧道。

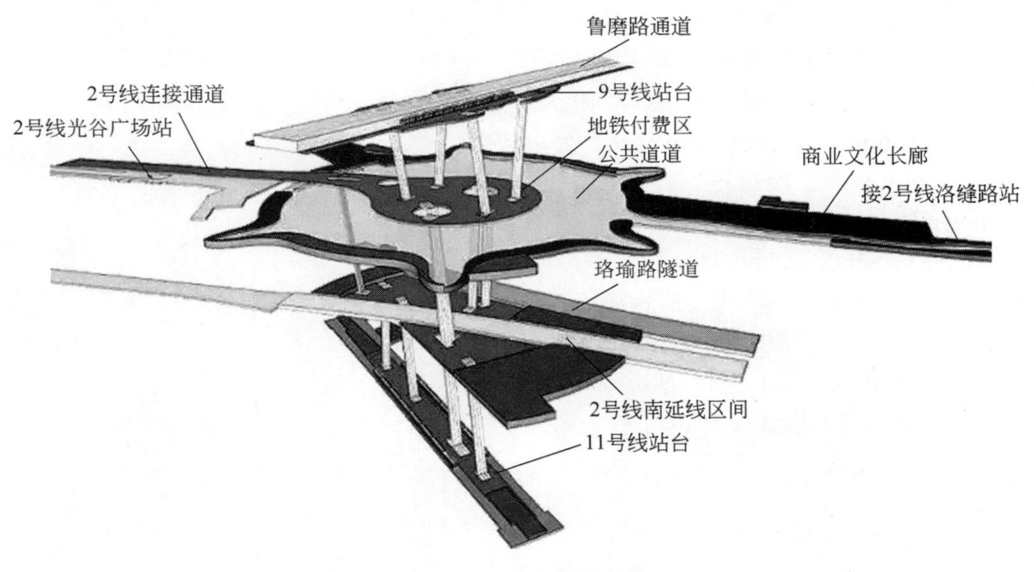

图7.7 光谷广场综合体

(Figure 7.7 Optics Valley Plaza Complex)

2. 地下公共空间(Underground public spaces)

Such a large volume of underground complex is quite rare both at home and abroad, its various designs also belong to the world-class quality level, after completion, the underground space is expected to reach 400000 person-times per day, this will effectively relieve traffic pressure and as a special bond to achieve regional commercial interworking.

其体量之大，国内外十分罕见，各类设计也属世界一流质量水平，建成后的地下空间人流预计可达日均40万人次，这将有效缓解交通压力并作为纽带实现地区商业互通。

7.4 专业词汇(Specialized Vocabulary)

geotechnical Investigations 岩土工程调查
alignment n. 线型，线路
soldier pile 支护桩
earth-retaining structure 挡土墙结构
shoring system 支撑系统
excavation n. 挖掘，发掘
earth-pressure balance (EPB) TBM 土压平衡掘进机
slurry-face TBMs 泥水平衡掘进机
in-line booster pump 直列式升压泵
integrated pipe gallery 综合管廊
underground pipe gallery 地下管廊
congestion n. 拥挤，堵塞
municipal facility 市政工程设施
earthquake-proof adj. 抗(地)震的
underground utility corridor 地下公用管廊
inspection well 检查井
cut-and-cover 明挖
tunnel project 隧道项目
shield-tunneling method 盾构法
shield machine 盾构机
pipe-jacking method 顶管法
hydraulic jack 液压千斤顶
corridor n. 廊(道)，管廊
trunk line 干线
branch line 支线
tank structure 液仓结构
cable duct gallery 缆线管廊
cover plate 盖板
hand hole 手孔
underground complex 地下综合体
natural hazard 自然灾害
high-rise building 高层建筑(物)
underground public space 地下公共空间

习题(Exercises)

1. Translate the following sentences into Chinese.

(1) The results of these investigations will influence final design and construction methods for stations, tunnels, other underground structures, and foundations.

(2) The EPB TBMs rely on balancing the thrust pressure of the machine against the soil and water pressures from the ground being excavated.

(3) Underground pipe gallery system not only solves the problem of urban traffic congestion, but also greatly facilitates the maintenance and overhaul of municipal facilities such as electricity, communication, gas, water supply and drainage.

(4) The cable duct gallery is generally located under the pavement of the road and is shallow.

2. Translate the following sentences into English.

(1) 由于综合管道内管线布置紧凑合理,因此有效利用了道路下的空间,节约了城市用地。

(2) 明挖法具有施工简单、快捷、经济、安全的优点,城市地下隧道式工程发展初期都将其作为首选的开挖技术。

(3) 地下综合体又称城市地下综合体,通常指可综合体现城市功能的大型城市地下空间。

(4) 修建地下车站时常采用“明挖法”。

防灾与减灾 (Disaster Prevention and Mitigation)

8.1 地震(Earthquake)

8.1.1 地震定义(The definition of earthquake)

An earthquake (also known as tremor) is the shaking of the surface of the Earth resulting from a sudden release of energy in the Earth's lithosphere that creates seismic waves. Earthquakes can range in size from those that are so weak that they cannot be felt to those violent enough to propel objects and people into the air, and wreak destruction across entire cities.

地震(又称震颤)是由地球岩石圈中能量突然释放产生的地震波所造成的地表震动。地震的规模有大有小,从微弱得感觉不到,到强烈得足以把物体和人身抛向空中,浩劫横扫整座城市。

At the Earth's surface, earthquakes manifest themselves by shaking and displacing or disrupting the ground. When the epicenter of a large earthquake is located offshore, the seabed may be displaced sufficiently to cause a tsunami. Earthquakes can also trigger landslides and occasionally, volcanic activity (Figure 8.1).

在地表上,地震表现为地层的晃动、移位或断裂。当大地震震中位于近海时,海床位移可以大到足以引起海啸。地震还能引发滑坡,偶尔也诱发火山爆发(图 8.1)。

The word earthquake is used to describe any seismic event—whether natural or caused by humans—that generates seismic waves (Figure 8.2). Earthquakes are caused mostly by rupture of geological faults but also by other events such as volcanic activity, landslides, mine blasts, and nuclear tests. An earthquake's point of initial rupture is called its hypocenter or focus.

地震一词用来描述任何地震性事件——无论是自然的还是人为的——产生地震性波动(图 8.2)。地震大多是由地质断层的断裂引起,也有其他因素所致,例如火山活动、滑坡、矿区爆破和核试验。地震的初始破裂点称为震中或震源。

Earthquake is a natural phenomenon that occurs without warning and does not respect cities, nations or borders. It is a judge of the efficiency of the antiseismic measures that are

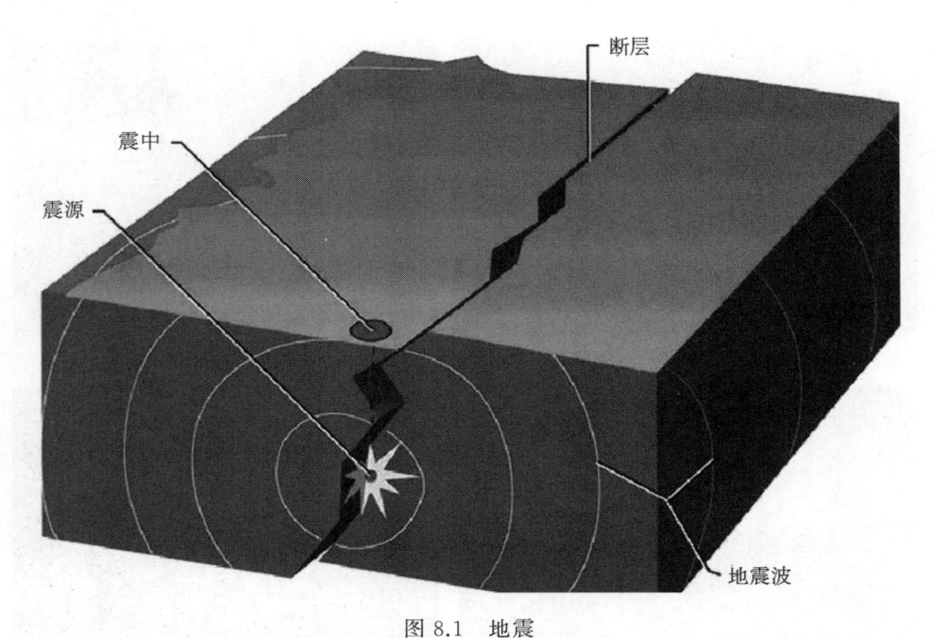

图 8.1 地震
(Figure 8.1 Earthquake)

applied to each construction, to the planning of a city and to the state level. When a large earthquake occurs, mistakes and incautiousness which were hidden under glamorous building facades are revealed and are, unfortunately, often accompanied by loss of human life (Figure 8.3).

地震是一种自然现象,发生时没有预兆,也不分城市、国家或边境。它是对于每一个建筑、城市规划和州一级层面的抗震性措施都适用的有效评判。大地震发生时,隐藏在华丽建筑表面下的错误和粗心就会暴露出来,不幸的是,往往伴随着生命的丧失(图 8.3)。

On the other hand, the earthquake is a message of life, a sign that our planet is and will remain alive, keeping mankind alive as well. This is because the earthquakes are the result of the continuous deformation and movement of the earth due to endogenous forces caused by gravity, rotation, and heat from within the Planet, and also the extraterrestrial forces coming from sources like solar radiation and the gravitational pull of the Moon and the Sun causing the tides etc. If it weren't for the endogenous forces that cause the earthquakes, the surface of the Earth would be covered by water of equal depth at all points. Therefore, the earthquake is indeed a message of liveliness for our Earth.

另一方面,地震是(地球)生命的一种信息,是我们地球"活着"的迹象,也是人类活着的标志。这是因为地震是地球由于重力、自转和来自地球内部热量所产生的内生力不断变形和运动的结果。也还有来自太阳辐射以及日月引力引起潮汐之类的地外力量。假如没有引发地震的内源性力量,地表就会被处处等深的水所覆盖。因此,地震确实是地球充满活力的信息。

图 8.2 地震受灾图

(a) 洛杉矶西北部卡里佐平原圣安德烈亚斯断层；(b) 2004 年印度洋地震引发的海啸；(c) 1906 年旧金山地震的大火；(d) 2010 年 1 月海地太子港的受损房屋

(Figure 8.2 Earthquake disaster map)

(a) The San Andreas Fault in the Carrizo Plain, northwest of Los Angeles; (b) The tsunami of the 2004 Indian Ocean earthquake; (c) Fires of the 1906 San Francisco earthquake; (d) Damaged buildings in Port-au-Prince, Haiti, January 2010

8.1.2 地震成因(The cause of the earthquake)

Earthquakes are usually caused when rock underground suddenly breaks along a fault. This sudden release of energy causes the seismic waves that make the ground shake. When two blocks of rock or two plates are rubbing against each other, they stick a little. They don't just slide smoothly; the rocks catch on each other. The rocks are still pushing against each other, but not moving. After a while, the rocks break because of all the pressure that's built up. When the rocks break, the earthquake occurs. During the earthquake and afterward, the plates or blocks of rock start moving, and they continue to move until they get stuck again. The spot underground where the rock breaks is called the focus of the earthquake. The place right above the focus (on top of the ground) is called the epicenter of the earthquake (Figure 8.4).

地震通常是在地下岩石沿着断层突然破裂时发生的。这种能量的突然释放会导致地震

图 8.3　1908 年 12 月 28 日，墨西拿地震和海啸

(Figure 8.3　The Messina earthquake and tsunami on December 28, 1908)

波，使地面震动。当两块岩石或两块板相互摩擦时，它们会有一点粘连。不只是滑动不再顺畅；它们会互相接合。仍然互相挤压，但并不移动。再过一会儿，岩石因压力全部聚集起来而破裂。岩石破裂时，地震就发生了。震时和震后，岩块或岩层开始移动，并且持续移动，直至它们再次被卡住。岩石破裂的地下点称为地震震源。震源正上方位置(地层顶部)称为地震震中(图 8.4)。

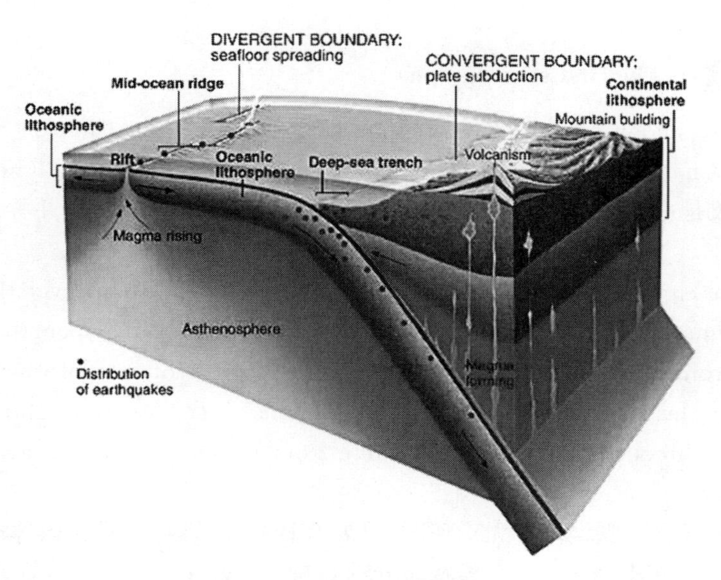

图 8.4　地震形成的概念图

(Figure 8.4　A concept map of the earthquake forming)

Earthquake-like seismic waves can also be caused by explosions underground. These explosions may be set off to break rock while making tunnels for roads, railroads, subways, or mines. These explosions, however, don't cause very strong seismic waves. You may not even feel them. Sometimes seismic waves occur when the roof or walls of a mine collapse. These can sometimes be felt by people near the mine. The largest underground explosions, from tests of nuclear warheads (bombs), can create seismic waves very much like large earthquakes. This fact has been exploited as a means to enforce the global nuclear test ban, because no nuclear warhead can be detonated on earth without producing such seismic waves.

地下爆炸也能引起类似地震的地震波。在为道路、铁路、地铁或矿区修建隧道时,这些爆炸可能会引发岩石破裂。然而,这些爆炸并不引起非常强烈的地震波。你甚至感觉不到。有时,当矿井的顶板或矿壁坍塌时,就发生地震波。矿区附近的人有时能感觉到。核弹头(炸弹)试验造成的最大地下爆炸能产生与大型地震非常相似的地震波。这一现象作为实施全球禁止核试验的一种手段,因为在地球上引爆任何核弹头都能产生这种地震波。

8.1.3 如何研究地震(How to study earthquakes)

Seismologists study earthquakes by going out and looking at the damage caused by the earthquakes and by using seismographs. A seismograph is an instrument that records the shaking of the earth's surface caused by seismic waves. The term seismometer is also used to refer to the same device, and the two terms are often used interchangeably.

地震学家通过外出观察地震所造成的损坏并采用地震仪来研究地震。地震仪是一种记录地震波引起的地表震动的仪器。地震计这一术语也用来指同一装置,这两个术语经常互换使用。

1. 首台地震仪(The first seismograph)

The first seismograph was invented in 132 A. D. by the Chinese astronomer and mathematician Chang Heng. He called it an "earthquake weathercock" (Figure 8.5).

首台地震仪是由中国天文学家兼数学家张衡于公元 132 年发明,并称之为"候风地动仪"(图 8.5)。

Each of the eight dragons had a bronze ball in its mouth. Whenever there was even a slight earth tremor, a mechanism inside the seismograph would open the mouth of one dragon. The bronze ball would fall into the open mouth of one of the toads, making enough noise to alert someone that an earthquake had just happened. Imperial watchman could tell which direction the earthquake came from by seeing which dragon's mouth was empty.

八条龙口中各含一颗铜球。无论何时,只要有轻微球震颤,地震仪内部的机械装置就会打开一条龙的嘴。铜球会落入一只蟾蜍张开的口中,并发出足够响的声音来警告人们刚刚发生了地震。皇家侍从通过查看哪条龙口在空着,就会分辨出地震来自何方。

In 136 A. D. a Chinese scientist named Choke updated this meter and called it a

图 8.5 张衡原创候风地动仪大型模型

(Figure 8.5 A large-scale model of Zhang Heng's original earthquake weathercock)

"seismoscope." Columns of a viscous liquid were used in place of metal balls. The height to which the liquid was washed up the side of the vessel indicated the intensity and a line joining the points of maximum motion also denoted the direction of the tremor.

公元 136 年,一位名叫科勒的中国科学家更新了地动仪,并称之为"验震器"。用黏性液柱换掉金属球。液体所冲刷容器一侧的高度表示强度,连接最高冲刷点的线段则表示震颤的方向。

2. 现代地震仪(Modern seismographs)

Most seismographs today are electronic, but a basic seismograph is made of a drum with paper on it, a bar or spring with a hinge at one or both ends, a weight, and a pen. The one end of the bar or spring is bolted to a pole or metal box that is bolted to the ground. The weight is put on the other end of the bar and the pen is stuck to the weight. The drum with paper on it presses against the pen and turns constantly. When there is an earthquake, everything in the seismograph moves except the weight with the pen on it. As the drum and paper shake next to the pen, the pen makes squiggly lines on the paper, creating a record of the earthquake(Figure 8.6). This record made by the seismograph is called a seismogram.

现代地震仪大多是电子的,但基本构造都是一只卷着纸的滚筒、一条一端或两端带有铰链的短棒或弹簧,一块重物和一支笔。短棒或弹簧的一端用螺栓固定在地面的杆或金属盒上。重物置于棒的另一端,笔粘在重物上。带卷纸的滚筒压在笔上不停转动。有地震时,地震仪部件除持笔重物外,一切都在运转。当鼓和纸在笔旁边震动时,笔就会在纸上划出一些弯弯曲曲的线条,从而记录地震(图 8.6)。这种由地震仪所做的记录称为地震记录。

By studying the seismogram, the seismologist can tell how far away the earthquake was and how strong it was. This record doesn't tell the seismologist exactly where the

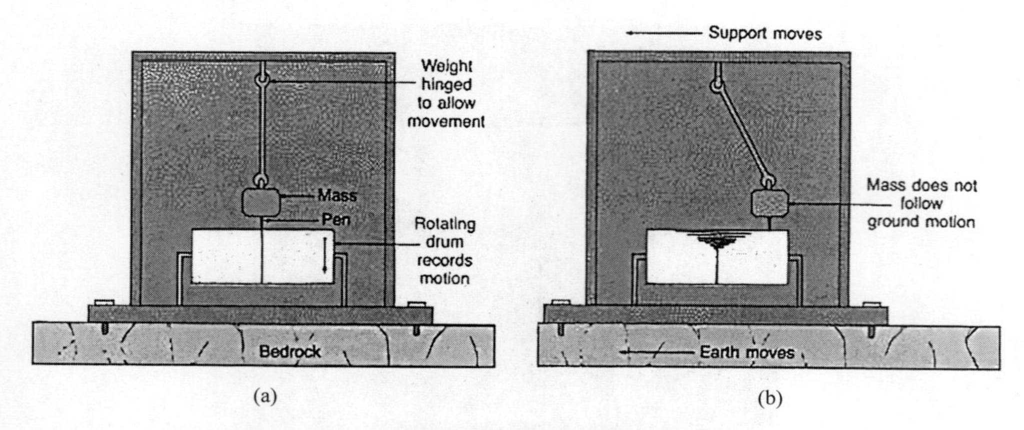

图 8.6 现代测震仪工作示意图

(Figure 8.6 Two illustrations of a modern seismograph in action)

epicenter was, just that the earthquake happened so many miles or kilometers away from that seismograph. To find the exact epicenter, you need to know what at least two other seismographs in other parts of the country or world recorded.

通过研究测震图,地震学家能辨别地震距此多远、强度多大。这种记录并不告诉这位地震学家震中的准确位置,只告诉他们发生地震距此测震仪多远。为了找到准确震中,至少需要知晓在指定国家或全球其他地方两台测震仪记录了什么。

8.1.4 地震预防(Precautions against earthquakes)

The things to pay attention to after the earthquake are as follows:

震后注意事项如下:

(1) Identify safe places in each room.

(1) 确定每一个房间的安全位置。

Under furniture such as a sturdy table; away from where glass could shatter—around windows, mirrors, pictures, or where book-cases or other heavy furniture could fall over.

在结实的桌子等家具下面;离开玻璃可能破碎之处——窗户,镜面、画框周围,或书架和其他沉重的家具会翻倒之处。

(2) Locate safe places outdoors.

(2) 设置户外安全区。

In the open, away from buildings, trees, telephone and electrical lines, overpasses, or elevated expressways.

在开阔处,与建筑物、树木、电话、电力线缆、立交桥和高架高速公路保持一定距离。

(3) Make sure all family members know how to respond after an earthquake.

(3) 确保所有家庭成员都知晓震后如何应对。

Teach all family members how and when to turn off gas, electricity, and water. Teach children how and when to call emergency services, the police, or fire department and which radio station to tune to for emergency information.

教会家人们如何以及何时关断煤气、电、水。教会儿童如何以及何时呼叫紧急服务、警察或消防部门，以及应该调到哪个广播电台收听应急信息。

(4) Contact local emergency management office or Red Cross chapter for information on earthquakes.

(4) 联系当地应急管理办公室或红十字会以获取有关地震的信息。

(5) Have disaster supplies on hand.

(5) 备有救灾补给品。

Flashlight and extra batteries, portable, battery-operated radio and extra batteries. First aid kit and manual, emergency food and water, non-electric can opener, essential medicines, sturdy shoes.

手电筒和备用电池、便携式电池供电的收音机及其备份电池、急救箱和手册、应急食物和水、非电动开罐器、必备药品、结实的鞋子。

(6) If outdoors.

(6) 在户外时。

Move into the open, away from buildings, street lights, and utility wires.

移至开阔处，与建筑物、路灯和公用设施线缆保持一定距离。

(7) If in a moving vehicle.

(7) 如果在行驶的车辆上。

Stop quickly and stay in the vehicle. Move to a clear area away from buildings, trees, over passes, or utility wires. Once the shaking has stopped, proceed with caution. Avoid bridges or ramps that might have been damaged by the quake.

迅速停车并留在车内。移至远离建筑物、树木、立交桥和电线的空地上。一旦震颤停止，谨慎前行。避开有可能被该次地震破坏的桥梁或坡道。

(8) Be prepared for aftershocks.

(8) 防备余震。

Although smaller than the main shock, aftershocks cause additional damage. Aftershocks can occur in the first hours, days, weeks, or even months after the quake.

余震虽比主震小，但造成附加损坏。余震能发生在震后的几小时、几天、几周甚至几个月。

(9) Help injured or trapped persons.

(9) 救助受伤或被困人员。

Give first aid where appropriate. Do not move seriously injured persons unless they are in immediate danger of further injury. Remember to help your neighbors, infants, the elderly, and people with disabilities,whoever may require special assistance.

在适当的地方给予急救。请勿移动受伤人员，除非他们处于紧急状态，否则他们会有二次受伤的危险。请记住要帮助可能需要特殊帮助的邻居、婴儿、老人、儿童和残疾人等。

Therefore, it is up to us, individual citizens, civil society and the state, to plan ahead and prepare for the unwelcome but inevitable future events, especially in Greece which has such a long experience.

因此,它取决于我们每位公民、社会、国家,事先筹划并准备好面对这些不想要但又无可避免的未来事件,尤其是在有着悠久历史的希腊。

8.2 塌方和突水(Collapse and Water Inrush)

8.2.1 塌方(Collapse)

During the construction ofhighway tunnels, the shear strength of the surrounding rock is reduced due to secondary stress adjustment, and the surrounding rock or artificial support structure loses its self-stabilizing and self-supporting capacity, resulting in the deformation and instability of the surrounding rock within a certain range of the tunnel. The following two situations have a higher probability of collapse. One is the moment when the excavation surface is blasted. Even the best controlled blasting will have a certain disturbance to the surrounding rock, and this disturbance just happens to be in the limit equilibrium state of the blast area or around the blast area. An additional load is applied to the surrounding rock, and this load just breaks this balance, which causes a collapse. Another case is the large-section tunnel, which is constructed in blocks and layers, and the excavation is completed at the lowest block-instantaneously it is easy to induce collapse.

塌方是指公路隧道施工期间,因二次应力调整降低了围岩抗剪强度,围岩或人工支护结构遂失去自稳自承能力,以致隧道一定范围内围岩变形和失稳。下面两种情况发生塌方的概率较大:一种是开挖面爆破瞬间,控制再好的爆破也会对围岩有一定扰动,而这恰好使爆区或其周围处于极限平衡状态。对围岩施加一种附加荷载,正是这种荷载打破了这一平衡,于是引发塌方。另一种情况是大断面隧道,分块分层施工,在最下部块体开挖完成瞬间容易诱发塌方。

8.2.2 塌方原因(The cause of collapse)

1. 地质因素(Geological factors)

Surrounding rock strength: Collapse mostly occurs in strong weathering zone, strong unloading relaxation zone, fault breaking and its intersection zone, and extremely weak rock layer. The lower the surrounding rock strength, the larger the scale of the collapse.

围岩强度:塌方大多发生在强烈风化带、强烈卸荷松弛带、断层破碎及其交汇带、极软弱岩层。围岩强度越低,塌方规模越大。

Rock mass structure: Compared with massive, layered and mosaic fragmented structures, bulk structures and fragmented structures are most likely to collapse.

岩体结构:同大块、层状及镶嵌碎裂结构相比,散体和碎裂状结构最可能发生塌方。

Unfavorable combination of structural planes: The unfavorable combination of structural planes is the main factor of collapse in layered and block structures. The wedge-shaped body composed of a herringbone cut in the top arch and a steep dip in the side wall is prone to instability.

结构面不利组合：在层状和块状结构中，结构面的不利组合是塌方的主要因素，由顶部的人字形切口和边墙的陡倾角所组成的楔形体容易失稳。

Groundwater: In hard rock, the stability of surrounding rock is not greatly affected by groundwater, while in soft rock, groundwater is the fatal factor of tunnel collapse.

地下水：硬岩中围岩稳定性受地下水影响不大，而软岩中地下水是隧洞塌方的致命因素。

In situ stress: As a general environment, in areas where vertical *in situ* stress is dominant, the tunnel arch is prone to instability, and in areas where horizontal *in situ stress* is predominant, tunnel side walls are prone to instability.

原地应力：作为普通环境，垂直原地应力主导的地区，隧道拱容易失稳，而以水平原地应力主导的地区，隧道边墙容易失稳。

2. 设计因素(Design factors)

Support type: Insufficient depth of exploration in the early stage or insufficient understanding of the rock and soil, resulting in a design support type that is inconsistent with the quality of the surrounding rock and provides insufficient support resistance, which leads to collapse.

支护类型：早期勘探深度不够或对岩土认识不足，导致设计支护类型与围岩质量不符，提供的支护抗力不够，因而引发塌方。

Support safety margin: the length of the anchor rod is too short to pass through the surrounding rock loose ring or anchor hole, the row distance is too large to form an effective pressure arch, and the cross-sectional size of the large pipe shed and arch frame is too small, so that the rigidity is weak. Waiting for it will bring instability of surrounding rock deformation.

支护安全裕度：锚杆长度偏短穿不过围岩松动圈或锚孔，排距偏大而未形成有效压力拱，以及大管棚和拱架的断面尺寸偏小以致刚度偏弱。这些问题不解决就会带来围岩变形失稳。

3. 施工因素(Construction factors)

Initial support: due to material supply, organizational coordination and other reasons, the surrounding rock is not supported in time after it is exposed, allowing the surrounding rock to relax and deform and eventually bring the surrounding rock to collapse. For tunnels excavated in layers and blocks, it is particularly important to emphasize that the initial support should be closed quickly.

初期支护：因材料供应、组织协调和其他原因，围岩揭露后未及时支护，任其松弛、变形，最终带来塌方。分层、分块开挖的隧道，特别强调的是初期支护宜快速闭合。

Excavation method: increase the excavation cross-section or blasting footage without authorization, so that the inherent stability of the surrounding rock will quickly generate excessive bending and shear stress to form a landslide.

开挖方法：擅自加大开挖面或爆破开挖，会使围岩因固有稳定性快速产生过多弯剪应力而形成滑坡。

Drilling and blasting: Drilling, charging, and connecting lines are not designed according to the blasting design, so that the single-shot or one-time initiation charge is too large, and the blasting vibration speed exceeds the specification (the specification requires no more than 15cm/s), which causes excessive damage to the surrounding rock, and a thick slack loop is formed, thereby inducing collapse.

钻孔与爆破：不按爆破设计表钻孔、装药、连线，使单响或一次起爆药量过大，爆破震动速度超规范要求(规范要求不大于15cm/s)，使围岩损伤过大，并形成厚松弛圈，由此诱发塌方。

Excavation quality: The excavation rock surface around the tunnel is uneven, and the undulation difference is obviously beyond the standard, which will cause a significant concentration of stress, which is obviously harmful to the stability of the cave.

开挖质量：隧洞周边的开挖岩面凹凸不平，起伏差明显超标，这会导致应力显著集中，对洞室稳定性显然不利。

Quality of initial support (including auxiliary measures): The quality of the initial support is fatal to the stability of the tunnel. The angle, length, grouting compactness of large and small pipe sheds and bolts do not meet the design requirements, the steel frame support is not closely attached to the surrounding rock of the tunnel, and the thickness and strength of the spray concrete are insufficient. These may weaken the initial support to the surrounding rock relaxing resistance.

初期支护(包括辅助措施)质量：初期支护质量对巷道的稳定性至关重要。大小管棚及锚杆的角度、长度、注浆密实度与设计要求不符，钢架支撑未与隧洞围岩密贴，喷射混凝土厚度、强度不足，这些都可削弱初期支护对围岩松弛的抗力。

8.2.3 突水(Water inrush)

In the process of highway tunnel construction, the groundwater in the surrounding rock aquifer overcomes the resistance of water-barrier layer, fault and fracture zone under the combined action of head pressure and other pressures, and then rushes into the tunnel in a sudden way, which is called water inrush. If groundwater carries large amounts of sediment or saturated sediment and suddenly rushes into the tunnel, it is called mud inrush. Mud inrush can be regarded as a derivative phenomenon of water inrush. Therefore, the key to the mud inrush is still water gushing. Without groundwater as the power and carrier, the mud process cannot occur. In general, the combination of high pressure, rich water and bad geology is the geological conductor to induce water inrush (mud).

公路隧道施工过程中，围岩含水层中的地下水在水头压力和其他压力的组合作用下，克服隔水层、断层、裂隙带的阻力，突然涌入隧道时，称为突水。当地下水携带大量泥沙或饱和泥沙并突然涌入隧道时，称为突泥。突泥被视为突水的衍生现象。因此，突泥的关键仍是涌水。不以地下水作为动力和载体，突泥就不能发生。总之，高压、富水、不良地质三者的结合，是诱发突水(泥)的地质原因。

8.2.4 突水成因(The cause of water inrush)

1. 含水围岩储能条件(Energy storage conditions of water-bearing surrounding rocks)

The energy storage conditions in which tunnel water inrush occurs refers to geological conditions that can form a large amount of groundwater and sediment. Karst, various broken zones in rock masses, and syncline structural basins and other parts have good water-enrichment and storage properties, and can often form groundwater bodies with large water volume and high water pressure. These parts are often also a source of abundant loose group material, or originally contain a large amount of loose group material. At the same time, these parts are also good exchange channels for groundwater. When the conditions are available, the groundwater in it or other groundwater bodies with hydraulic connection are poured into the tunnel together with mud and stones.

隧道突水发生的储能条件指能形成大量地下水及泥沙的地质状况。岩溶、岩体中的各种破碎带以及向斜结构盆地等部位，具有良好的富水和储水性能，常能形成水量大、水压高的地下水体。这些部位往往也是松散团材料的丰富来源，或本为含有大量的松散团材料。同时这些部位也是地下水的良好交换通道，在条件允许的情况下，将隧道内的地下水或其他有水力连接的地下水与泥、石块一起涌入隧道。

2. 含水围岩释能条件(Energy release conditions of water-bearing surrounding rocks)

A large amount of energy is stored in the water-bearing surrounding rocks, but whether tunnel water inrush can occur depends on the tunnel energy release conditions, that is, the main conditions for controlling the tunnel water inrush are its energy release conditions, including water pressure and relative water-barrier thickness.

含水围岩中储存了大量能量，但隧道突水能否发生，还取决于隧道释能条件，即控制隧道突水的主要条件就是其释能条件，包括水压及相对隔水层厚度。

Rock masses with different rock mass structures and lithological conditions have large differences in physical and mechanical properties, and the thickness of the water-resisting layer is also different. The minimum amount of water inrush required for different types of surrounding rock to be breached by water inrush is shown in the Table 8.1.

不同岩体结构和岩性条件的岩体，物理力学性质差异很大，阻水层厚度也不同。不同类型围岩被突水击破所需的最小突水量见表 8.1。

表 8.1 不同类型围岩被涌水突破所需的最小突水量

(Table 8.1 The amount of small water burst required by water inrush for different types of surrounding rock)

Rock category (岩石类别)	Limestone (石灰岩)	Clastic (碎屑岩)	Metamorphic rock (变质岩)	Clay rock (黏土岩)
Minimum water inrush(最小突水量)	100～200m/h^3	125～300m/h^3	150～200m/h^3	10～40m/h^3

3. 含水围岩的稳定性(Stability of water-bearing surrounding rocks)

Tunnel excavation directly affects the stability of the water-bearing surrounding rock and causes water inrush in the tunnel. If the opposite water-repellent layer is directly excavated, the exposed groundwater body will produce an instant water burst. Even if there is a certain thickness of water-repellent layer at the palm face, due to construction blasting, or the surrounding rock relaxation and stress concentration caused by tunnel excavation, the surrounding rock will be deformed and destroyed, which will reduce the effective protection thickness of the relative water-repellent layer thereby increasing the possibility of tunnel water inrush.

隧道开挖直接影响到含水围岩的稳定性,造成隧道突水。当直接开挖掉对立的驱水层时,裸露的地下水体会瞬间爆发。即使掌子面处存在一定厚度的驱水层,由于施工爆破,或隧道开挖带来的围岩松弛和应力集中,围岩将发生变形破坏,这会使相对驱水层的有效保护厚度减少从而增加隧道突水的可能性。

8.2.5 隧道作业中的危险(Hazards in tunnel operations)

All the health and safety hazards of normal civil engineering construction can be found in tunneling along with a few which are specific to tunneling. In most cases the risks arising from these hazards present more severe consequences in tunneling. This increase in severity is due to a number of factors including:

正常土木工程施工中,所有的健康和安全隐患都能在隧道作业中找到,也有一些是隧道作业特有的。大多数情况下,这些危害引起的风险在隧道作业中带来更严重后果。这种严重性的增加受多种因素影响,包括:

(1) The degree of uncertainty in the nature and variability of the ground through which the tunnel is being driven.

(1) 隧道掘进作业中所通过处的地层的自然性质和可变性的不确定性程度。

(2) The confined space of the tunnel environment particularly in small utility tunnels.

(2) 隧道环境中的受限空间,特别是在小型公用设施隧道中。

(3) A safety culture at all levels in the workforce which has until recently been poorly developed.

(3) 员工队伍中的各层级人员至今却没有培养出较好的安全意识。

(4) A lack of commitment from all parties to the project in addressing occupational health and safety.

(4) 项目各方缺乏对职业健康和安全的承诺。

(5) Failure by the industry, to learn from the experiences and mistakes of others.

(5) 行业失败,要从他人的经验和错误中学习。

(6) Work in compressed air.

(6) 在压缩空气环境中工作。

8.3 滑坡和泥石流(Landslides and Mudslides)

8.3.1 滑坡(Landslides)

Landslides occur when masses of rock, earth or debris move down a slope. Mudslides, also known as debris flows or mudflows, are a common type of fast-moving landslide that tends to flow in channels.

当大量的岩石、泥土或碎屑沿着斜坡下移时,就发生滑坡。泥石流,又称碎屑流或泥浆流,是一种普通的快速移动滑坡类型,往往在沟槽中流动。

8.3.2 滑坡原因(The cause of the landslides)

Landslides are caused by disturbances in the natural stability of a slope. They can happen after heavy rains, droughts, earthquakes or volcanic eruptions. Mudslides develop when water rapidly collects in the ground and results in a surge of water-soaked rock, earth and debris. Mudslides usually begin on steep slopes and can be triggered by natural disasters. Areas where wildfires or construction have destroyed vegetation on slopes are at high-risk landslides during and after heavy rains (Figure 8.7).

滑坡由斜坡自然稳定性的扰动引起。它们可能发生在暴雨、干旱、地震或火山爆发之后。当水迅速积聚在地面,造成了浸水的岩石、泥土和碎屑激增时,就会发生泥石流。泥石流通常始于陡峭的斜坡上,可能由自然灾害引发。被野火或建筑破坏了山坡植被的地区在暴雨过后是高风险的滑坡地区(图 8.7)。

(a)

(b)

图 8.7　滑坡和泥石流

(a) 滑坡;(b) 泥石流

(Figure 8.7　Landslides and Mudslides)

(a) Landslides; (b) Mudslides

In the United States, landslides and mudslides result in 25 to 50 deaths each year. The health hazards associated with landslides and mudslides include:

在美国,滑坡和泥石流每年导致 25～50 人死亡。与滑坡和泥石流相关的健康危害包括:

(1) Rapidly moving water and debris that can lead to trauma.

(1) 快速流动的水和碎屑能导致创伤。

(2) Broken electrical, water, gas and sewage lines that can result in injury or illness.

(2) 破损的电气、水、煤气和排污的管道能导致损伤或疾病。

(3) Disrupted roadways and railways that can endanger motorists and disrupt transport.

(3) 道路和铁路的中断,可能危及机动车驾车员并扰乱运输。

So what's the difference between a mudslide and a landslide?

那么,泥石流和滑坡有什么差别?

Landslides are the movement of rock and debris down a slope. The majority of landslides are caused by multiple conditions acting in unison to destabilize a slope. Wildfires, heavy rainfall, earthquakes, environmental degradation, and volcanic eruptions are known to trigger landslides.

滑坡是指岩石和碎石沿斜坡向下移动。大多数滑坡是由多重条件联合作用而使斜坡失稳引起的。众所周知,野火、暴雨、地震、环境恶化和火山喷发都会引发滑坡。

8.3.3 泥石流(Mudslides)

Mudslides, also known as debris flows, are a specific type of landslide where the debris flows in rapid channels. They typically occur when soil on steep slopes becomes so saturated with rain that any friction dramatically decreases and the slope collapses due to the influence of gravity. These are typically more destructive due to their rapid nature.

泥石流,又称碎屑流,是一种特定类型的滑坡,其中泥石在快速沟渠中流动。它们一般发生在陡坡上的土壤,因雨水饱和使得摩擦力急剧降低,并因重力作用而致斜坡坍塌。由于其快速的性质,通常更具破坏力。

In the case of the Oso mudslide, the area had experienced increased levels of rainfall: nearly 15 inches above anticipated levels. Unauthorized clear cutting of forests on the hillside exacerbated the situation, allowing water to accumulate under the soil until gravity overcame friction (Figure 8.8(a)).

在奥索泥石流的例子中,该地区历经降雨量不断增加:比预期高出近 15 英寸。山坡上未经授权的砍伐森林恶化了这种情势,使水到土壤层下聚集,直到重力大过摩擦力(图 8.8(a))。

The mudslide in Montecito, Calif., was caused by an influx of rain over areas engulfed by the Thomas Fire that burned over 281 000 acres in Ventura and Santa Barbara counties. The fire destroyed most of the vegetation that would have otherwise soaked up the rainwater and made the terrain less susceptible to mudslides (Figure 8.8(b)).

加利福尼亚州蒙特西托泥石流是由于雨水大量涌入了托马斯大火所吞噬的地区。这场大火烧毁了文图拉和圣巴巴拉两县超过 281 000 英亩的土地,摧毁了原本会吸收雨水的大部分植被,使得地形不易受泥石流影响(图 8.8(b))。

A slow-moving landslide on the western slope of Rattlesnake Ridge in Union Gap, Wash. has forced dozens to evacuate as officials and residents prepare for the seemingly inevitable collapse of the ridge. Over 70 GPS monitors and three seismometers have been

(a)

(b)

图 8.8 奥索泥石流与加利福尼亚州蒙特西托泥石流
(a) 奥索泥石流；(b) 加利福尼亚州蒙特西托泥石流
(Figure 8.8 Oso mudslide and the mudslide in Montecito, Calif)
(a) Oso mudslide; (b) The mudslide in Montecito, Calif

installed throughout the area of the landslide to track its movement and provide an advanced warning of the looming threat. Last October, the landslide was recorded moving "less than an inch per day". Now, it's moving up to 2.5in. per day (Figure 8.9).

在华盛顿州联合峡(Union Gap)的响尾蛇山脊(Rattlesnake Ridge)西坡上，缓慢移动的滑坡迫使数十人撤离，政府官员和居民准备应对山脊上不可避免的坍塌。已安装了 70 多台 GPS 监视器和 3 台地震仪遍布滑坡区，用来跟踪滑坡的运动并为迫在眉睫的威胁提供预警。2016 年 10 月，滑坡记录为"日均不到 1 英寸"。而现在上升到日均 2.5 英寸(图 8.9)。

图 8.9 截至 2017 年 12 月 31 日，响尾蛇岭上的裂缝
(Figure 8.9 The fissure on Rattlesnake Ridge as of Dec. 31, 2017)

8.3.4 泥石流成因(The cause of the mudslides)

Debris flows can be triggered by intense rainfall or snowmelt, by dam-break or glacial outburst floods, or by landsliding that may or may not be associated with intense rain or earthquakes. Debris flows can be more frequent following forest and brush fires, as experience in southern California demonstrates(Figure 8.10).

强降雨或融雪会引发泥石流,决堤或冰川暴发洪水也会,或与强降雨和地震有关联的滑坡。正如南加州的经验所表明,随森林和灌木丛火灾之后,泥石流可能会更加频繁(图 8.10)。

图 8.10　加利福尼亚州静泉垭口的古泥石流沉积

(Figure 8.10　Ancient debris flow deposit at Resting Springs Pass, California)

Debris flows are accelerated downhill by gravity and tend to follow steep mountain channels that debouche onto alluvial fans or floodplains. The front, or "head" of a debris-flow surge often contains an abundance of coarse material such as boulders and logs that impart a great deal of friction. Trailing behind the high-friction flow head is a lower-friction, mostly liquefied flow body that contains a higher percentage of sand, silt and clay. These fine sediments help retain high pore-fluid pressures that enhance debris-flow mobility. In some cases, the flow body is followed by a more watery tail that transitions into a hyperconcentrated stream flow. Debris flows tend to move in a series of pulses, or discrete surges, wherein each pulse or surge has a distinctive head, body and tail.

泥石流在重力作用下加速下坡,往往沿陡峭山道流向冲积扇或泛滥平原。在泥石流的前部或"头部",常含有大量粗糙材料,例如巨石和原木,两者产生很大的摩擦力。在高摩擦"头部"的后面,是一种低摩擦、大多经液化的流动体,含有较高比例的沙子、粉砂和黏土。这些细小的沉积物有助于保持高孔隙流体压力,从而增强泥石流的流动性。在某些情况下,流体后面是一个更富水的"尾部",过渡到一个高度集中的水流。泥石流倾向于以一系列脉冲或离散的涌浪移动在那里,每一脉冲或涌浪都有独特的"头部"、"身体"和"尾部"。

Some areas are more likely to experience landslides or mudslides, including:

一些地区更容易发生滑坡或泥石流,包括:

(1) Areas where wildfires or construction have destroyed vegetation.

(1) 被森林大火或建筑已破坏了植被的地区。

(2) Areas where landslides have occurred before.

(2) 以前曾发生过滑坡的地区。

(3) Steep slopes and areas at the bottom of slopes or canyons.

(3) 陡峭斜坡和斜坡或峡谷底部的区域。

(4) Slopes that have been altered for construction of buildings and roads.

(4) 因建筑物和道路的施工而改变的斜坡。

(5) Channels along a stream or river.

(5) 沿溪流或河流的沟渠。

(6) Areas where surface runoff is directed.

(6) 地表径流流向的区域。

(7) Things to note after landslides and mudslides occur

(7) 发生滑坡和泥石流后的注意事项

(8) Stay away from the site. Flooding or additional slides may occur after a landslide or mudslide.

(8) 离开该地区一定距离。滑坡和泥石流之后可能会发生洪水或其他滑坡。

(9) Check for injured or trapped people near the affected area, if it is possible to do so without entering the path of the landslide or mudslide.

(9) 在不进入滑坡和泥石流路径的情况下,检查受影响区域附近的受伤或被困人员。

(10) Listen to the radio or TV for emergency information.

(10) 收听广播或电视以获悉应急信息。

(11) Report broken utility lines to the appropriate authorities.

(11) 向适当部门报告已破损的公用设施线缆。

(12) Consult a geotechnical expert for advice on reducing additional landslide problems and risks. Local authorities should be able to tell you how to contact a geotechnical expert.

(12) 向岩土工程专家咨询有关减少额外滑坡问题和风险的建议。当地有关部门应该告知住户如何能够联系上岩土工程专家。

8.4 专业词汇(Specialized Vocabulary)

tremor n. 震颤

the Earth's lithosphere 地球岩石圈

seismic wave 地震波

landslide n. 滑坡

volcanic adj. 火山(引起)的

natural adj. 天然的

rupture n. 断裂

geological adj. 地质(学)的

mine blast 矿区爆破

nuclear test 核试验

planet n. 星球
endogenous force 内生力
building n. 建筑(物)
aftershock n. 余震
mudslide n. 泥石流
slope n. 斜坡
volcanic eruption 火山喷发
debris n. 碎屑
debris flow 碎屑流
gravity n. 重力
friction n. 摩擦;摩擦力
collapse n. 塌陷
GPS monitor GPS监视器
seismometer n. 地震计
geotechnical adj. 岩土(工程)的

习题(Exercises)

1. Translate the following sentences into Chinese.

(1) An earthquake (also known as tremor) is the shaking of the surface of the Earth resulting from a sudden release of energy in the Earth's lithosphere that creates seismic waves.

(2) Earthquake is a natural phenomenon that occurs without warning and does not respect cities, nations or borders.

(3) Landslides occur when masses of rock, earth or debris move down a slope.

(4) The majority of landslides are caused by multiple conditions acting in unison to destabilize a slope.

(5) Rock mass structure: Compared with massive, layered and mosaic fragmented structures, bulk structures and fragmented structures are most likely to collapse.

(6) Rock masses with different rock mass structures and lithological conditions have large differences in physical and mechanical properties, and the thickness of the water-resisting layer is also different.

2. Translate the following sentences into English.

(1) 当大地震的震中位于海上时,海床可能发生位移而引起海啸。

(2) 地震主要是由地质断裂引起的,也有其他因素引起的,例如火山活动、滑坡、爆炸和核试验。

(3) 滑坡是由斜坡自然稳定性的扰动引起的。

(4) 在暴雨期间,因被山火或建筑而破坏了斜坡上植被的地区处于高风险滑坡之中。

(5) 隧道突水发生的储能条件指能形成大量地下水及泥沙的地质条件

(6) 隧道开挖直接影响到含水围岩的稳定性,造成隧道突水。

参考文献(References)

[1] 李亚东.土木工程专业英语[M]. 成都：西南交通大学出版社，2005.

[2] 方旭明. 新编专业外语[M]. 成都：西南交通大学出版社，1997.

[3] 文军，杨全红，贺武. 专门用途英语教程——以学习为中心的方法[M]. 重庆：重庆大学出版社，1996.

[4] REESE L C, ISENHOWER W M, WANG S T. Analysis and design of shallow and deep foundations [M]. New York: John Wiley & Sons, Inc., 2006.

[5] AWOSHIKA K, REESE L C. Analysis of foundations with widely spaced batter piles[C]// Proceedings. the international symposium on the engineering prop-erties of seafloor soils and their geophysical identification, University of Washington, Seattle, 1971.

[6] BIENIAWSKI Z T. Rock mechanics design in mining and tunneling[M]. Balkema: Rotterdam/ Boston, 1984.

[7] DONALD P C. Geotechnical engineering: principles and practices[M]. India: Prentice Hall of India, 2002.

[8] BERNHARD M, LEONHARD S, WILL R, et al. Hardrock tunnel boring machines[M]. Germany: Deutsche Nationalbibliothek, 2008.

[9] GARY B. Practical tunnel construction[M]. New Jersey: John Wiley & Sons, Inc., Hoboken, Inc, 2013.

[10] FABIO B, PAOLO C, MARCO D, et al. Road tunnel an analytical model for risk analysis[M]. Switzerland: Springer Nature Switzerland AG, 2019.

[11] MARCO D, BOSCARDIN. Jacked tunnel design and construction[M]. Boston: Geotechnical Special Publication, 1998.

[12] DIMITRIOS Kolymbas. A rational approach to tunnelling[M]. Berlin: Tunnelling and Tunnel Mechanics, 2008.

[13] CUI Z D, ZHANG Z L, ZHAN Z X, et al. Dynamics of freezing-thawing soil around subway shield tunnels[M]. Singapore: Springer Nature Singapore Pte Ltd, 2020.

[14] PUHAKKA T. Underground drilling and loading handbook[M]. Tampere Finland: Tamrock Corp, 1997.

[15] HARMAN, H L. Introductory mining engineering[M]. 2nd ed. New York: John Wiley & Sons, Inc, 2002.

[16] DE LA VERGNE J. Hard rock miner's handbook[M]. 3rd ed. Tempe/North Bay: McIntosh Engineering, 2003.

[17] MORRISON P R, MCNAMARA A M, ROBERTS T O, et al. Design and construction of a deep shaft for corssrail[J]. Geotechnical Engineering, 2004, 157(4): 173-182.

[18] WEN F, MOU S, OUYANG M, et al. Design and construction of shaft-driving type piezoceramic ultrasonic motor[J]. Ultrasonics, 2004, 43(1): 35-47.

[19] WARD C R. Analysis and significance of mineral matter in coal seams[J]. International Journal of Coal Geology, 2002, 50(1): 135-168.

[20] BROUGH B H. A rudimentary treatise on coat and coal mining[J]. Nature, 1900, 61(1583): 411-412.

[21] ROSS M, ROTH M. All nine alive: the story of the quecreek mine rescue[J]. Pittsburgh Post-

Gazette，2002，4.

[22] GRIFFITHS D V，LANE P A. Slope stability analysis by finite elements[J]. Geotechnique，1999，49(3)：387-403.

[23] WANG L，HUANG J H，JIANG C H，et al. Reliability-based design of rock slopes—a new perspective on design robustness[J]. Engineering Geology，2013，154：56-63.

[24] AKI K. The use of love waves for the study of earthquake mechanism[J]. Journal of Geophysical Research，1960，65(1)：323-331.

[25] SCHOLZ C H，SYKES L R，AGGARWAL Y P，et al. Earthquake prediction：a physical basis[J]. Science，1973，181(4102)：803-810.

答案(Answers)

第2章

1.

(1) 深入全面的现场勘察是土木工程设计和施工必不可少的环节。

(2) 基础施工的开挖程序需要清理场地、放线、开挖及根据开挖深度采取安全措施。

(3) 基础开挖布置时,应在现场用砌石柱建立基准,并与就近的标准基准连接。

(4) 墙的中心线应延长,并在柱子的石灰顶上标明。

(5) 基础工程完成后,在收回木构件时要小心,以免沟槽坍塌。

2.

(1) A detailed account of the nature, quantity, availability and significant properties of materials considered for construction purposes.

(2) Before the excavation for the proposed foundation is commenced, the site shall be cleared of vegetation, brushwood, stumps of trees etc.

(3) The center lines of the perpendicular walls are marked by setting out the right angle with steel tapes or preferably witha theodolite.

(4) For small buildings, excavation is carried out manually by means of pickaxes, crow bars, spades etc.

(5) The exact spacing can be decided on the basis of the type of soil.

第3章

1.

(1) 如果地基的深度小于或等于地基的宽度,通常则认为是浅地基。

(2) 地基必须稳定,以抵抗支承土的剪切失效。

(3) 地基沉降不能超过容限范围,以免损坏结构。

(4) 结构荷载可以通过桩传递到更深的坚硬地层。

(5) 每一基础的设计对工程师都是一次独特的挑战。

2.

(1) Obtain the required information concerning the nature of the superstructure and the loads to be transmitted to the foundation.

(2) Piles may be classified as long or short in accordance with the L/d ratio of the pile (where, $L=$ length, d =diameter of pile).

(3) To create a foundation for good performance, the engineer must carefully address the topics and other related factors in the above list.

(4) The total settlement of a structure is the maximum amount the structure has settled with respect to its original position.

第4章

1.

(1) 由于具备大的适用范围,钻爆法对可变地质状况的施工是比较有利的。

(2) 力通过该钻杆传递给钻头,然后侵入岩石。钻头根据冲击钻进、旋转钻进或者两者兼有的钻进方式设计成不同种类。

(3) 钻头冲击或锤击可产生通过压应力和剪应力共同来破碎岩石。

(4) 波从钻孔传播到最近的自由面(此距离称为负载),并反射回钻孔。

(5) 敞开式盾构主要用于强度高且稳定的围岩,但也有某些类型的掘进机可用于软土地层。

2.

(1) Although a majority of tunnel construction occurs within the ground, for example TBM and NATM tunnelling, there are techniques, most notably immersed tube tunnels, which are constructed differently.

(2) The rock is cut using these excavation tools by the rotation of the cutter head and the blade pressure on the face.

(3) This compressive shock wave travels from the borehole though the entire rock mass as an elastic wave, with its velocity, a function of rock density; the denser the rock, the faster the wave travels.

(4) Generally the main differences are the cutter tools, also known as cutter dressing, on the cutter head and the face support requirements as the ground stability is generally lower.

第5章

1.

(1) 一级和二级水平巷道在矿山开发中起着重要作用。

(2) 这在统计数据中得到了很好的证明,1980年所开发井总长度中,水平井占82%,上升井占16%,其余为竖井、斜井等所占。

(3) 斜井矿山不同于竖井矿山和平巷矿山,前者通过垂直向下的隧道掘进,后者通过水平方向的隧道掘进。

(4) 边坡矿井的排水通风可以利用主边坡进行,也可以利用副井或钻孔进行。

(5) 地下采矿作业的设计要求运输、通风、地面控制和采矿方法相结合,形成一个系统,为矿山人员提供最高程度的安全。

2.

(1) Shallow shafts, typically sunk for civil engineering projects differ greatly in execution method from deep shafts, typically sunk for mining projects.

(2) Today shaft sinking contractors are concentrated in Canada, Germany and South Africa.

(3) Slope mining is a method of accessing valuable geological material, such as coal

or ore.

(4) The size of timbers required depends upon the load they are to bear.

(5) The second grave concern was the quality of the air in the mine.

第6章

1.

(1) 对于自然边坡和人工边坡,地质、边坡材料性质和地形始终是重要因素。

(2) 自然边坡是自然界中存在、由自然原因形成的边坡。

(3) 变形通常发生在斜坡内的主要自然不连续面、古滑动面和构造剪切带上。

(4) 涉及压实土的填方边坡包括铁路和公路路堤、土坝和堤坝。

(5) 边坡稳定性是指被土壤覆盖的斜坡承受和经受运动的潜力。

(6) 在城市环境中,几乎所有沿峡谷而下的步道都存在斜坡稳定性问题。

2.

(1) The slope is one of the most basic natural geological environments for human survival and the project activities.

(2) The stress history of the slope materials is of tremendous importance.

(3) Man-made slopes are formed by humans as per requirements.

(4) The slopes formed by unnatural process.

(5) It is often necessary to consider the stability of an embankment-foundation system rather than that of an embankment alone.

(6) When it has been established that a slope is potentially unstable, reinforcement may be an effective method of improving the factor of safety.

第7章

1.

(1) 地质勘查的结果将影响到车站、隧道和其他地下建筑以及地基工程的最终设计方案和施工方法。

(2) EPB TBM 依靠平衡机器的推力压力与挖掘地面的土体和水压力从地面进行挖掘。

(3) 地下综合管廊系统不仅解决了城市交通拥堵问题,还大大方便了电力、通信、燃气、供排水等市政设施的维护和检修。

(4) 缆线管廊一般设置在道路人行道下面,其埋深较浅。

2.

(1) Due to the compact and reasonable layout of pipelines within the comprehensive pipeline, the space under the road is effectively utilized and urban land is saved.

(2) Open excavation has the advantages of simple, fast, economical and safe construction, and it is the preferred excavation technology in the early stage of urban underground tunnel project development.

(3) Underground complexes, also known as underground urban complexes, usually refer to large-scale urban underground spaces that can comprehensively embody urban

functions.

(4) Construction of the underground stations would employ the cut-and-cover construction technique.

第8章

1.

(1) 地震(又称震颤)是由地球岩石圈中能量突然释放而产生的地震波所引起的地球表面的震动。

(2) 地震是一种自然现象,发生时没有预兆,也不分城市、国家或边界。

(3) 当大量的岩石、泥土或碎屑沿斜坡滑下时,就会发生滑坡。

(4) 大多数滑坡是由多种条件共同作用所导致斜坡失稳造成的。

(5) 岩体结构:同块状、层状及镶嵌碎裂结构相比,散体结构、碎裂状结构更容易发生塌方。

(6) 不同岩体结构和岩性条件的岩体,物理力学性质存在较大差异,隔水层厚度也不尽相同。

2.

(1) When the epicenter of a large earthquake is located offshore, the seabed may be displaced sufficiently to cause a tsunami.

(2) Earthquakes are caused mostly by rupture of geological faults but also by other events such as volcanic activity, landslides, mine blasts, and nuclear tests.

(3) Landslides are caused by disturbances in the natural stability of a slope.

(4) Areas where wildfires or construction have destroyed vegetation on slopes are at high-risk landslides during and after heavy rains.

(5) The energy storage conditions in which tunnel water inrush occurs refers to geological conditions that can form a large amount of groundwater and sediment.

(6) Tunnel excavation directly affects the stability of the water-bearing surrounding rock and causes water inrush in the tunnel.